T0129335

SOLUTIONS TO THE MYSTERY OF CONSCIOUSNESS

TAPAN DAS

SOLUTIONS TO THE MYSTERY OF CONSCIOUSNESS

iUniverse books may be ordered through booksellers or by contacting:

iUniverse
1663 Liberty Drive
Bloomington, IN 47403
www.iuniverse.com
1-800-Authors (1-800-288-4677)

ISBN: 978-1-5320-5220-0 (sc)
ISBN: 978-1-5320-5219-4 (e)

Library of Congress Control Number: 2018907166

Print information available on the last page.

iUniverse rev. date: 07/11/2018

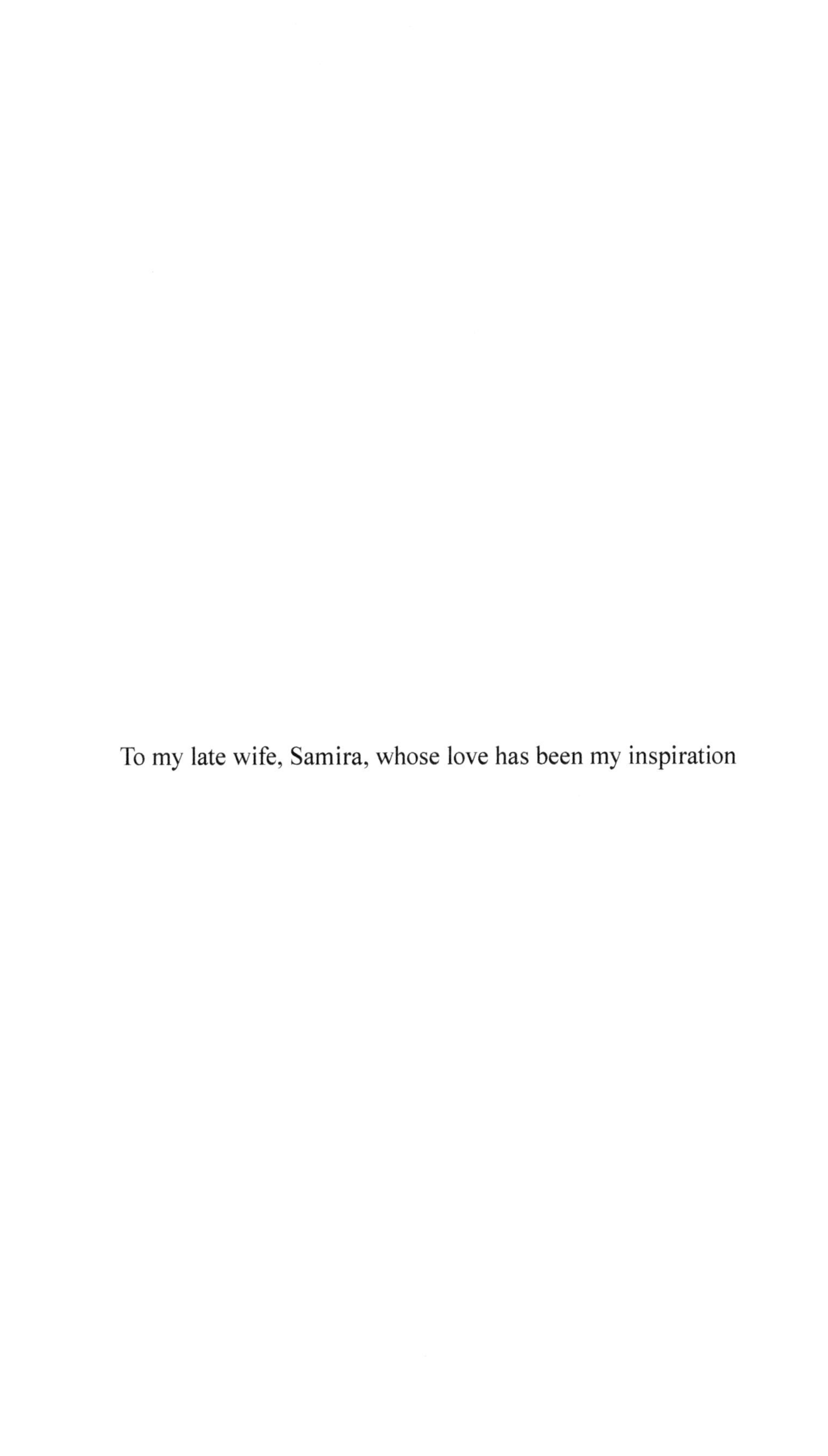

To my late wife, Samira, whose love has been my inspiration

CONTENTS

PREFACE

My family is Hindu. My parents were very religious. My father always had Bhagavad Gita, divine discourse spoken by the Supreme Lord Krishna to warrior Arjuna as a true source of spiritual knowledge revealing the purpose and goal of human life. Hinduism strongly believes in dualism. Atman, or soul, in the human body borrows consciousness from the all-pervading source Brahman, or Divinity. As I grew older and started studying science and engineering, I was increasingly questioning this dualistic view of consciousness since there is no scientific proof of it. However, being extremely busy in my working life, I could not pursue my interest in consciousness. After retiring and publishing a book, *Why Astrology Is Science: Five Good Reasons*, I focused totally on researching consciousness. I published three papers on consciousness based on quantum brain dynamics. I was still not convinced of the origin of consciousness.

I then started studying the structure and function of the human brain. I found that the frequency of the brain waves depends on the alertness of mind—that is, the level of consciousness. Hence, consciousness is a function of brain wave frequency. My new theory was published in *Neuroquantology* (September 2017, volume 15, issue 3) with good reviews.

This book is targeted for all types of audience. Consciousness has been a subject of interest since early civilization. Many great minds, including those of scholars and scientists, have worked on it. However, no common or scientific agreement has yet been reached. Monumental work is currently being carried out by neuroscientists to understand the function of our brain and consciousness. My theory of consciousness as a property of brain and as a function of brain wave frequency would help them to solve the mystery of consciousness.

ACKNOWLEDGMENTS

When my wife, Samira, was seriously sick with liver cancer and was in palliative care, we moved to Ottawa from Mississauga, Canada, to live with our daughter, Neeta. Although Neeta took exceptional care of Samira, Samira passed away on September 28, 2014. After recovering from her death, I continued my research on consciousness. I sincerely thank Neeta for taking very good care of me, supported by her husband, Angus. My thanks also go to Dr. Ray Barton, CEO of Vitesse, and Mr. David Carter of Ottawa University for helping me with a few diagrams.

INTRODUCTION

Consciousness has been a mystery since the early civilization of humankind. Consciousness has been defined as the awareness of one's own existence, sensations, thoughts, surroundings, and so forth. Philosophers in every religion and culture have struggled to comprehend the nature of consciousness and identify its essential properties. After rapid developments in science and technology over the past few decades, consciousness has become a significant topic of interdisciplinary research in cognitive science, with significant contributions from fields such as psychology, neuropsychology, neuroscience, and quantum mechanics.

The word *conscious* was originally derived from the Latin word *conscius* (*con* meaning "together" and *scio* meaning "to know"), meaning "joint or common knowledge with another." The Latin phrase *conscius sibi*, meaning "conscious unto oneself," is more closely related to the current concept of consciousness. *Consciousness* should not be confused with *conscience*, which has the more moral connotation of knowing when one has done, or is doing, something wrong. Through consciousness, one can have knowledge of the external world or one's own mental states. Sometimes consciousness is simply meant to refer to the fact that an organism is awake, as opposed to sleeping or being in a coma. However, state of consciousness is often implied by creature consciousness—that is, the organism is having conscious mental states.

Since early days, every religion, philosophy, and philosopher has discussed consciousness. In Hinduism, consciousness is prana, which flows through every channel of the body and keeps us alive. Hence, Upanishads describe it as the soul of the body and equate it with atman or the essence of Brahman. In Buddhism, it is *vijñāna* (Sanskrit) or *viññāṇa* (Pali), translated as consciousness, meaning "sense cognition" or "sense awareness." Human consciousness in Chinese philosophy may be seen in three layers: cosmological consciousness, consciousness of the human self, and political consciousness. In Taoist thought, humans are centered not in a brain but in the heart and mind. Consciousness is a statement of both inner and outer connections. In Greek philosophy, Pythagoras and Plato accepted the preexistence of the "nous," or divine soul of humanity, which chooses the existence for which it must incarnate. It survives the death of the body, and if it has not attained sufficient perfection to merit endless bliss, it must be subjected to new tests by reincarnating in order to attain further progress and perfection. In Mayan civilization, consciousness has been defined as a pyramid of seven levels, from awareness of physical body to principles of honesty, faithfulness, service, and truthfulness. For the ancient Egyptians, *ba* animated a living person, whereas *ka* was the energy emanating from that person, which might refer to spirit and soul. Another important aspect of Egyptian belief represented immortality, the ankh, depicted as the crested ibis. Consciousness in early religions and civilizations is discussed in detail in chapter 1.

As civilization progressed and science became a dominant factor of our society, the concept of consciousness changed. Currently, there are two scientific views of consciousness: monistic and dualistic. Monistic view is that the brain does all the functions; as such, there is nothing called consciousness. Dualistic view is that brain and consciousness are two separate things. Monistic view considers the brain a supercomputer that can do all the necessary functions. I discuss monism and dualism in detail, with views of different religions, scientists, and philosophers, in chapter 2.

With the advent of computer and artificial intelligence, scientists and engineers are passionately working on creating consciousness in artificial intelligence. I have discussed this in detail in chapter 3.

I have discussed in chapter 4 several well-known scientific theories of consciousness: integrated information theory, global workspace theory, McGinn's theory, David Chalmers's theory, PANIC theory, higher-order monitoring theory, Rosenthal's higher-order thought theory, cognitive theory, and views of other researchers.

Quantum mechanics changed all concepts of Newton's classical physics. It explained the structure of atoms to the level of electron, proton and neutron, wave-particle duality, uncertainty principle, and so on. Quantum consciousness became an important area of research of many scientists. Electrons can tunnel between adjacent neurons in the brain, thereby creating a virtual neural network. This virtual neural network produces consciousness. I have discussed quantum consciousness in detail in chapter 5.

Cosmology has been a subject of interest from since early civilization. Twinkling stars and the sun, moon, and sky have intrigued people to find the source and its creation. While astronomy deals with each individual object in the universe, cosmology deals with the creation, evolution, and composition of the universe. In many religions and philosophies, universal consciousness considers consciousness part of the universe. I have discussed this in chapter 6.

Consciousness acts differently with time than with shape or size. This is subjective time. Different levels of consciousness may experience time in different ways. I have discussed this in chapter 7.

I have worked extensively in quantum consciousness and published three papers. However, not being satisfied with this approach, I have created a new theory and finally concluded that consciousness is a property of the brain. It is a function of brain waves and physical constant *Conscire C*. Different types of brain waves with different frequencies are created in the brain depending on the states of consciousness. This constant *Conscire C* is like gravitational constant, Planck constant, and so forth, and it is multiplied by the brain wave frequencies to create consciousness. This has been discussed in detail in chapter 8. I hope that this theory will solve the mystery of consciousness.

CHAPTER 1

Consciousness in Early Religions and Civilizations

This chapter will discuss in detail how consciousness was viewed and described by different religions, early civilizations, and philosophers. Consciousness has been a mystery, and all religions and philosophers have tried to explain it in different ways. The mystery has been how body, mind, and spirit interact and what happens after death. Does consciousness die with death? Does the spirit stay after death? If the spirit stays, where does it go? Let us now see how different religions and philosophers view it.

Hinduism and Hindu Philosophy

According to Hinduism, consciousness is spirit. The spirit has its existence independently in Divinity, within and beyond creation. Ancient rishis (seers or sages) taught that the mind can expand to such a degree that it gradually begins to become one with its former self. The energy of the soul actually merges with the energy of the universe. As the mind expands more and more, the practitioner enters a state known as samadhi. From there, the yogi can eventually leave the body entirely, feeling the universe as his body.

The Vedas (Sanskrit), meaning knowledge, are large body-of-knowledge texts originating in the ancient Indian subcontinent. The

Vedas are considered revelations seen by ancient sages after intense meditation, and the texts have been carefully preserved since ancient times. There are four Vedas: the Rigveda, the Samaveda, the Yajurveda, and the Atharvaveda. Each Veda has been subclassified into four major text types: the Samhitas (mantras and benedictions), the Aranyakas (text on rituals, ceremonies, sacrifices, and symbolic sacrifices), the Brahmanas (commentaries on rituals, ceremonies, and sacrifices), and the Upanishads (texts discussing meditation, philosophy, and spiritual knowledge). The Rigveda (Sanskrit: from *ṛc* [i.e., praise] and *veda* [i.e., knowledge]) is a collection of 1,028 hymns that contain the mythology of the Hindu gods. It is considered one of the foundations of the Hindu religion and was composed in the northwestern region of the Indian subcontinent, most likely between 1500 and 1200 BC. The Samaveda (Sanskrit: *sāman* [song] and *veda* [knowledge]) is the Veda of melodies and chants. The Yajurveda (Sanskrit: *yajus* [mantra] and *veda* [knowledge]) is the Veda of prose mantras. The Atharvaveda (Sanskrit: *atharvāṇ* [procedures of everyday life] and *veda* (knowledge) is the knowledge storehouse of Atharvāṇ, a legendary Vedic sage. It gives the procedures for everyday life.

Upanishads are philosophical and different from religious aspects of Vedas. Upanishads written in the Indian subcontinent between 800 BC and 500 BC are different from the traditional Vedic religious order, dealing mainly with internal spiritual quests. There are two hundred surviving Upanishads, but fourteen are considered the most important Upanishads. Each of them is complete in itself.

Upanishad has four basic principles: samsara, karma, dharma, and moksha. Samsara is the cycle of death and rebirth in life. Karma is action in one's life that determines fate in future existence. Dharma is dutiful service to society. Moksha is the liberation from cycle of rebirth by the law of karma. Atman is a person's soul, and Brahman is the ultimate reality of one's existence. When a person achieves moksha, atman returns to Brahman. Most Indian religions, including Hinduism, Jainism, and Buddhism, share philosophical principles of samsara, karma, dharma, and moksha.

There are four states of consciousness described in Upanishad: *jagrat*, *svapna*, *susupti,* and *turiya*.

1. Jagrat means "awaking." We are aware of our daily world in this state.
2. Svapna means "dream." This is the inner subtle body.
3. Susupti means "deep sleep." In this state, consciousness is undistracted.
4. Turiya means "pure consciousness." This transcends the other three common states of consciousness. In this consciousness, both Saguna Brahman and Nirguna Brahman are transcended. Saguna Brahman provides the true state of experience of the infinite (*Ananta*) and nondifferent (Advaita/*abheda*). Nirguna is the eternal, all-pervading, and omnipresent divine consciousness.

According to Hindu philosophy, consciousness is not the neuro-activities of the central nervous system as viewed by neuroscientists and mind is not a function or process created by the brain. Mind is the inner instrument, or *antahkarana* in Sanskrit. This inner instrument becomes conscious by borrowing consciousness from the only source that is Brahman, or Divinity. Brahman is all-pervading. It is present behind everything and every mind-body complex as the foundation. Brahman is the very core of every being and is called the divine self, or the atman.

According to Hinduism, consciousness has two universal states existing in all creations. One is universal, eternal, and pure. The other is with qualities, states, and dynamism. The source of the first state is Brahman, which provides pure consciousness, and the source of the second state is prakriti, or nature, which provides self or ego consciousness—an aspect of nature. Ego consciousness (Sanskrit: *chit*) is the whole body and mind consciousness. Ego consciousness creates body consciousness infused with power and dynamism of nature. Pure consciousness, known as *satchitananda* (*sat* plus *chit* plus *ananda*), is infused with power of truth. In Sanskrit, *sat* means "truth," *chit* means "consciousness," and *ananda* means "bliss." Universal or pure consciousness is the center of human consciousness.

According to science, a single entity or singularity created the universe, maintaining and governing the fundamental machinery of everything in this universe. In Hinduism, Brahman, the one supreme and universal, is the origin and support of the universe. The first book of Hindus, named Rig Veda, proclaims, "Ekam Sat, Viprah Bahudha Vadanti," which means the following: "There is only one truth, only men describe it in different ways." According to Hindu philosophy, God and individuals are described as "anor aniyan mahato mahiyan," meaning "God is smaller than the smallest and greater than the greatest." This means that whether something is extremely large or infinitesimal, it is still made of the same divine source. God is omnipresent, meaning present everywhere and in everyone. According to quantum mechanics, at the subatomic level, there is no particle but waves. This coincides with the ideas of sunyata and maya, or illusion in Hindu philosophy, covering the whole universe: "Brahman Satyam, Jagat Mithya." Brahman is the only truth; the world is a false illusion. Because of the covering of maya, we do not see the underlying real Brahman. Similarly, in quantum mechanics, one sees only the material objects around and does not see strange quantum waves. According to theory of entanglements, two electrons demonstrate instant correlated properties even at distances where no communication is possible between them during the given time. This gives rise to the idea of the interconnected wholeness of the world, similar to Brahman in Hindu philosophy. Another finding of quantum theory is the involvement of the observer and the observed things according to Heisenberg's uncertainty principle. It is impossible to separate the effect of the measuring apparatus from the object measured. Detachment of the two is just not possible. Such an idea about the observer and the object of observation is also emphasized in Upanishads. It is behind the holistic philosophy about mind and body.

The Upanishads are a collection of texts of religious and philosophical nature, written in India probably between 800 BCE and 500 BCE, during a time when Indian society started to question the traditional Vedic religious order. The Chandogya Upanishad explains that all that exists is the self. It gives examples that a tall tree has its essence, the self, originally in the small seed from which it grew. Yet breaking a seed open will reveal no such potency for it to grow into such a huge plant. But the power is there. Likewise, when salt is mixed with water, it makes the salt invisible. However, by tasting the water, we know that the salt is there. Similarly, in the material body, the soul, or atman, exists, though we do not directly perceive it. Similarly, we also cannot see the soul in the body except by recognizing the symptom, which is consciousness. Brihadaranyaka Upanishad says that whoever is dear to us, whether it is our wives, husbands, sons, daughters, or so forth, is dear to us only due to the presence of the soul within the body. Once the soul leaves the body when a person is dead, the body becomes unattractive to us. Therefore, the body is not our real identity but the soul within.

Judaism and Jewish Philosophy

Judaism was founded over thirty-five hundred years ago in the Middle East. It encompasses the philosophy, religion, and culture of the Jewish people. Abraham, his son Yitshak (Isaac), and grandson Jacob (Israel) are referred to as the patriarchs of the Jewish people. The name Israel is derived from the name given to Jacob.

God promised Abraham that he would look after the Jews. But the Jews were living as slaves in Egypt. A prophet called Moses led the Jews out of slavery in Egypt and led them to the Holy Land that God had promised them. By parting the Red Sea, God helped Jews escape. When they reached Mount Sinai, God spoke to Moses and renewed the deal he had made with Abraham. God gave Moses the Torah, which contains statements of laws and ethics.

Sabbath is Judaism's day of rest and the seventh day of the week, when religious Jews and certain Christians (such as Seventh-day Adventists and Seventh-day Baptists) remember the biblical creation of the heavens and the earth in six days, as well as the exodus of the Hebrews.

In Hebrew, kavvanah is the concept that all our actions need to be done with conscious intention, focusing on actions and, in particular, *mitzvot*, or sacred actions. In Judaism, God is one. Conscious Judaism strives to enhance people's awareness, intentionality, and self-knowledge. It improves people's engagement with the world, their decision-making process, and how they view themselves. It helps people in pursuing *tikkun olam* (repairing the world) and *tikkun hanefesh* (repairing the soul).

Jewish philosophy includes all philosophy carried out by Jews in relation to the religion of Judaism. Athens and Jerusalem are the roots of Western civilization. From Jerusalem came Judaic monotheism and its expression in the Hebrew and Christian Bibles, which are the ultimate source of many of the metaphysical concepts underpinning the worldview, values, and sensibility of the Western mind. According to the book of Genesis, the world is the product of neither chance nor necessity; rather, it is created ex nihilo (out of nothing) in a deliberate act of will by God, who is transcendent to it and whose absolute unity is the ground of all existence. In the Hebrew Bible, God is supra-personal, not subpersonal; that is, God is not portrayed as a principle or as a material force or process but rather as a Being possessing intelligence, freedom, and will. Man shares in God's freedom and creativity, from which follows the sanctity of human life and the dignity of the human person. The first important Jewish philosopher was Philo Judaeus, a leader of Alexandrian Jewry, medieval rabbi, and philosopher. Moses Maimonides continued this process of attempting to reconcile reason and revelation, philosophy and religion, the Greek and the Hebrew. The major Jewish figures in philosophy after Moses were Baruch de Spinoza, Karl Marx, Edmund Husserl, Henri Bergson, Ludwig Wittgenstein, and Sir Karl Popper.

There are five levels of consciousness in Judaism. They are called (in ascending order) Nefesh, Ruach, Neshama, Chaya, and yechida.

> Nefesh: It is the awareness of the physical body and the physical world, the world of Asiya, the world of action. Nefesh is the life force of the body; hence, it has an awareness of the body.
>
> Ruach: Its primary manifestation is in the emotions.
>
> Neshama: Its primary activity is in the conceptual grasp of the intellect, and the soul. It has the notion of coming into being from nothingness, rather than structured, quantified existence. Thus, Neshama has the concept of continuous creation and sustenance of life and existence.

Chaya: It gazes upon the divine energy of the world. It communes with God as He transcends the worlds. This is the knowledge of the absolute truth of things.

Yechida: It corresponds to the level of soul called Adam Kadmon, the first spiritual world that came into being after the contraction of God's infinite light. This is the essence of the soul which is naturally and immutably bound to the Holy One.

Jainism

Jainism is an ancient world religion established by Mahavira in about 500 BCE in the Indian subcontinents. Jainism has some similarities with both Hinduism and Buddhism. It has its own unique and distinguishing features and is a truly ascetic religion which places heavy emphasis upon ethical conduct, personal purity for the liberation of the individual souls. Jainism does not believe in God. It believes that the universe is a self-existing system. Jainism believes that soul is eternal but that it has states, shapes, and sizes. In the bound state (*bandha*), it is subject to karma, resulting in rebirth. It attains liberation from rebirth only when it is able to get rid of karma completely. Jainism was founded by twenty-four perfect beings known as tirthankaras, or prophets, who appeared upon Earth in the past and laid the foundation for the doctrine of Jainism. Mahavira was the last and the most popular of the tirthankaras. Jainism enjoyed popularity only among certain sections of Indian society, especially merchant communities. It is a predominantly Indian religion, with some following abroad.

Life and consciousness are coextensive. Wherever there is life, there is consciousness, and vice versa. But there are degrees of explicitness or manifestation of consciousness in different organisms. In the lowest class of organisms, it is latent, while in human beings, it is manifest. Jiva is entirely distinct from inanimate existence, which does not possess consciousness.

Early Egyptian Philosophy

There is very little extant information on ancient Egyptian philosophy today. Egyptian society was based on the concept of *Ma'at*, which means balance and order. The ancient Egyptians believed that man was composed of three parts: the body, spirit, and soul. The fate of the soul was determined by its actions during life, whether good or bad. The first of the soul-like elements in the Egyptian system is the *ka*. It is a life force bestowed at birth. The ka needs to be considered together with the other main soul element, the ba. Although the ka is specific to each individual, the ba contains all the characteristics of the individual, especially the personality. Although both were to some degree dependent on the *khat* (body), the ba was tied to it rather more closely; the ka could wander off even during sleep and after death was the first to move on to a new afterlife. When the person successfully negotiated the judgment that awaited after death, the ba eventually joined the ka and they merged, forming the final immortal stage of the person, the akh.

Buddhism

Buddhism is an Indian religion attributed to the teachings of Buddha. Buddhism does not believe in God but believes in humanity. Gautama Buddha was the founder of the religion. He lived between 600 and 400 BCE. Gautama Buddha was said to be the prince of a little kingdom that was in modern Nepal. His father wanted him to be a great king and tried to keep his son from all religion and sights of death and suffering. When Buddha grew up, he was shocked by seeing an old man and a corpse. Then he wanted to solve suffering and death. He vowed to sit under a tree until he knew the truth, and he became enlightened when he was thirty-five. Then he started teaching. He taught that everybody could be enlightened. He contradicted the Hindu belief that only high-caste people might be holy, which threatened the hierarchical society. It is said that many disciples became arhats (worthy persons), and he taught everybody of all castes.

Pali is the language of the scriptures of Buddhism. The Tripitaka (called Tipitaka in Pali) is the earliest collection of Buddhist writings.

Initially they were composed orally but were written down by the third century BCE. The word means “the three baskets” (tri is three, *pitaka* is baskets) and refers to the way the texts were first recorded. The early writing material used consisted of long, narrow leaves sewn together on one side. Bunches of these were then stored in baskets. This is a large collection, running forty-five volumes in one modern edition.

Mind is another sense base in Buddhism. There are five external sense bases: eye, ear, nose, tongue, and body. Mind is internal sense base. Thus there are six internal and external sense bases. Internal sense base, or mind, gives rise to consciousness. The products of external sense organs—that is, image, sound, smell, taste, and touch—react with consciousness, creating feeling and craving. This is called Pali Canon’s Six Sextets, as shown in figure 1.

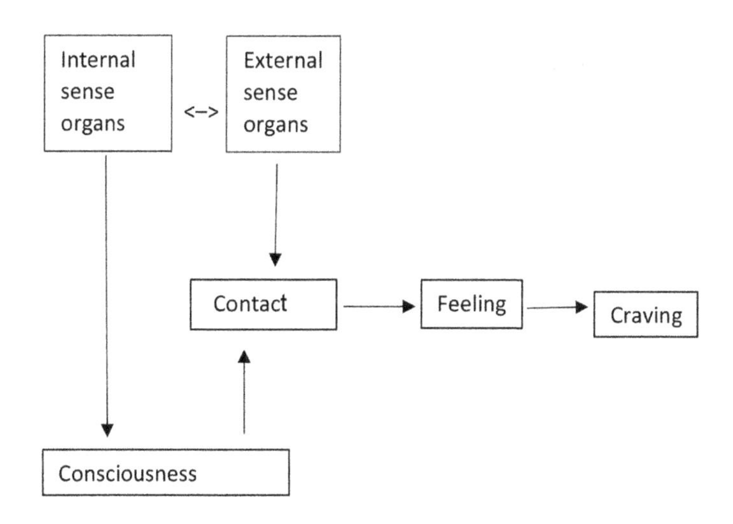

Figure 1: The Pali Canon’s sextets

More specifically, the six types of consciousness are eye consciousness (that is, consciousness based on the eye), ear consciousness, nose consciousness, tongue consciousness, body consciousness, and mind consciousness. When an ear’s sense organ receives a sound, the associated ear-related consciousness arises. The arising of the three elements—ear, sound, and ear consciousness—causes a pleasant, unpleasant, or neutral feeling to arise. It is from such a feeling that craving arises (figure 1).

According to Buddhism, the cleansing of one's mind of its defilements is vital. Defilements arise from craving, anger, and ignorance of not seeing and comprehending things in their authentic form. The cleansing is done with systematic forms of meditation that progress into knowledge, wisdom, and insight. With the cleansing of the mind, the kammic energy that flows for one runs its course to a finish. *Kamma*, or karma in Sanskrit, means action or doing. With rebirth, disease, decay, suffering, and death are repeated. When kammic energy ceases, rebirth ends. In Buddhism, mind and consciousness are the same. Consciousness with sensual faculties creates bodily activities, feelings, sensations, perceptions, and mental formations. Consciousness is the most vital part in the life of man. The Dalai Lama has said, "According to Buddhism, we cannot posit a beginning to consciousness. If we do so, then we would have to accept a first instance of consciousness that is uncaused and has come from nowhere." Buddhism believes that coming together of causes and conditions creates everything. So, causes create consciousness. Matter alone cannot produce consciousness. Consciousness is generated from a previous instance of consciousness, which is experience.

Christianity

Christianity is the world's largest religion, with over 2.4 billion followers, or 33 percent of the global population, known as Christians. Christianity is a monotheistic religion based on the life and teachings of Jesus Christ. Jesus was born in 6 BCE in Bethlehem. His mother, Mary, was a virgin who was betrothed to Joseph. The philosophy of Christianity is a way of life. It involves the reason for human life—why human life was created and sustained by God and nourished by His Spirit. We should all be following a good manner of life and setting our standards of conduct with Jesus Christ as the center stage. According to Christianity, human beings are children of God. Religious consciousness—the consciousness of religious truth—is the highest possible proof of the reality of religion. True religion is a matter of consciousness.

The primary distinctions between Christianity and Judaism include the identification of Jesus as the Son of God and the belief that Jesus died (and arose from the dead) as a final sacrifice for human sins.

Most Christians understand the soul as an ontological reality distinct from, yet integrally connected with the body. Its characteristics are described in moral, spiritual, and philosophical terms. When people die, their souls will be judged by God and determined to spend an eternity in heaven or in hell.

Roman Catholic Beliefs

The Catholic Church defines the soul as "the innermost aspect of humans, that which is of greatest value in them, that by which they are most especially in God's image. Soul signifies the spiritual principle in man. The Catholic Church teaches that the existence of each individual soul is dependent wholly upon God: "The doctrine of the faith affirms that the spiritual and immortal soul is created immediately by God."

Protestant Beliefs

Protestants generally believe in the soul's existence but fall into two major camps about its meaning in terms of afterlife. Some believe in the immortality of the soul and conscious existence after death, while others believe in the mortality of the soul and unconscious "sleep" until the resurrection of the dead.

Origin of Soul

The origin of soul has been provided by three major theories: soul creationism, traducianism, and preexistence. According to creationism, each individual soul is created directly by God, either at the moment of conception or at some later time. According to traducianism, soul is transmitted through natural generation along with the body, the material aspect of human beings. According to the preexistence theory, the soul exists before the moment of conception.

Chinese Philosophy

The main sources of Chinese philosophy were philosophers and schools that flourished from the sixth century BCE to third century BCE, an era of great cultural and intellectual expansion in China. This was called the golden age of Chinese philosophy because a broad range of philosophies and ideas were developed and discussed freely. These philosophies have profoundly influenced lifestyles and social consciousness even up to the present days in East Asian countries. The main schools of Chinese philosophy are Confucianism, Mohism, Taoism, Legalism, agrarianism, and Buddhism.

Confucianism

Confucianism was developed from the teachings of the Chinese philosopher Confucius (551–479 BCE). It is described as a philosophy, a religion, and a way of life. Confucianism regards the human activities and relationships as a manifestation of the sacred aspects of spirits or gods (*shén*). Confucianism rests on the belief that human beings are fundamentally good, teachable, improvable, and perfectible through self-cultivation and self-creation. Confucian thought focuses on the cultivation of virtue and maintenance of ethics.

Five Classics and Four Books are the authoritative books of Confucianism written before 300 BCE. The Five Classics are the Book of Documents, Book of Odes, Book of Rites, Book of Changes, and the Spring and Autumn Annals. The Book of Documents describes the history of ancient China, notably the Xia, Shang, and Zhou dynasties. The Book of Odes is the oldest existing collection of Chinese poetry, dating from the eleventh to seventh centuries BCE. The Book of Rites describes the ceremonial rites, the social administration, and social forms of the Zhou dynasty. The Book of

Changes deals with a system similar to geomancy, which interprets markings on the ground or the patterns formed by tossed sands, soil, or rocks. Spring and Autumn Annals are historical chronicles of the state of Lu. It was previously understood that Confucius wrote the annals. But now it is believed that they were written by various scholars in the state of Lu.

Four books are the Doctrine of the Mean, the Great Learning, Mencius, and the Analects. Doctrine of the Mean encourages people to engage in moral self-cultivation by following the principle of Tao. It emphasizes that rulers become effective by maintaining balance and encouraging Tao in people. The Great Learning teaches moral self-cultivation by learning. This keeps one in harmony, leading to consistent moral behavior. Mencius believes that human nature is basically good and strongly emphasizes the responsibility of the emperor to practice good governance. Analects believe that people attain virtue by learning. Analects also emphasize the importance of good governance, virtue, piety, and ritual.

From the Han dynasty (206–220 CE) to the early Song dynasty (960–1279 CE), the Five Classics grew into thirteen classics. In the early Song dynasty, scholars focused on the original Five Classics. By the mid–Song dynasty, however, the Analects, Mencius, Great Learning, and Doctrine of the Mean began gaining importance.

Confucianism explains the development of human consciousness as both individual and communal. This can be explained in terms of the *li* (ritual) and *ren* (deep feeling). Li serves to maintain a social order founded on family relations. Ren is the deep feeling of a person for the unity of humanity and a consciousness of an underlying bond among human persons that would lead to the love, care, and regard of one for the other. Confucius believed that the practice of li based on the feeling of ren would render li a meaningful and living force that would both regulate one inside and harmonize human relationships outside.

Taoism and Mohism

Taoism, also known as Daoism, was developed in fourth century BCE. It is a religious or philosophical tradition of Chinese origin that emphasizes living in harmony with the Tao (meaning "way"). The Tao is a fundamental origin by virtue of which all things happen or exist. Taoism differs from Confucianism by rejecting the practice of li (rituals) and social order. Tao is a source of being from which things rise and to which they will return. Taoism emphasizes wu wei (effortless action), naturalness, simplicity, spontaneity, and the three treasures: jing (energy of the physical body), qi (life force, including the thoughts and emotions), and shén (spirit or generative power). In Taoism, one practices the nonseparation between oneself and the Tao, where the Tao is to be embodied in all things in the world, large or small. The whole world is to be understood with an open and creative mind, which would link the outer world of nature to the inner world of human spirit.

Chi (pronounced chee) is the Chinese word used to describe the natural energy of the universe. This energy, though called natural, is spiritual or supernatural and is part of a metaphysical belief system. Chi is thought to permeate all things, including the human body. Chi is made up of two dynamic opposites: yang and yin. Yang is the masculine active principle in nature and is exhibited in light, heat, or dryness. Whereas yin is the feminine passive principle in nature, exhibited in darkness, cold, or wetness. The human body has both yang and yin chi aspects to it. If yang and yin are not in balance, the internal organs and other body systems progressively deteriorate and become diseased.

Figure 2: Taijitu, the yin-yang symbol of Taoism

The symbol of the Taijitu (figure 2) is meant to show the principles of yin and yang duality of Taoist philosophy. The outer circle symbolizes unity. The dark and light areas of the symbol indicate that within unity, there is duality of yang and yin, implying the combination of male/female, light/dark, excitable/quiet, and in/out. The small circles within the dark and light areas represent that there is nothing that is entirely separate from anything else. The center point indicates the balance point for all aspects of life.

Mohism originated in the teachings of Mo Di, or Mozi (Master Mo, 430 BCE), from whom it got its name. Mohism was an influential philosophical, social, and religious movement that flourished during the Warring States era (479–221 BCE) in ancient China. During that time, Mohism was seen as a major rival to Confucianism but later merged into Taoism, disappearing as an independent school of thought. Mohism sees Confucianism as dividing social classes into the ruled and the ruling, which lack a common base for solidarity. According to Mohism, social justice must be founded on one identical standard. Mozi diagnoses the cause of continuous wars as the lack of love among people. In order to reach peace among warring states, he advocates the doctrine of universal love (*jian-ai*). Love is universal (*jian*) if it can be shared on an equal basis. The ultimate ideal of Mohism is this: every person and every state would live on an equitable basis, and a society of human life would flourish just as a society of natural life flourishes under the compassionate will of heaven.

Legalism

Legalism was a philosophical belief in ancient China that human beings are more inclined to do wrong than right because they are motivated entirely by self-interest. Hence, social harmony can be achieved by strong state control and absolute obedience to authority. Qin dynasty (221–207 BCE) was brutally implementing this policy. However, this led to the dynasty's overthrow and end of Legalist philosophy in China.

Agrarianism

Agrarianism is a social philosophy that values rural society as superior to urban society, the independent farmer as superior to the paid worker. Farming is the sole occupation that offers total independence and self-sufficiency. It has the following tenets:

> Urban life, capitalism, and technology destroy independence, fostering vice and weakness.
>
> The agricultural community is the model society with its tenet of labor and cooperation.
>
> The farmer has a stable and solid family with sense of identity.
>
> The agricultural community is the model society with its fellowship of labor and cooperation.

Chinese Buddhism

Buddhist teachers may have arrived in the third century BC in China. Buddhist teachings started during the Tang dynasty (618–907 AD), and it became very popular and powerful. Buddhism in China merged with native Taoism and folk religion. Buddhism and Buddhists were sometimes supported and sponsored by the rulers during the past two thousand years. Nevertheless, sometimes Buddhists were eradicated and temples and scriptures were destroyed to make people not believe it. "Happy Buddha" is commonly seen in China and is depicted as being fat and laughing or smiling. The main goal of life in modern China is to be happy. Maybe that is why Buddha is shown this way. Happy Buddha has been the common popular Buddha in China for hundreds of years.

Zen Buddhism is a school of Mahayana Buddhism that originated in China during the Tang dynasty as Chan Buddhism. Zen school was strongly influenced by Taoism and developed as a distinguished school of Chinese Buddhism.

Japanese Philosophy

Japanese philosophy has been a fusion of both indigenous religion Shinto and continental religions such as Buddhism and Confucianism. It has been heavily influenced by both Chinese philosophy and Indian philosophy, as with Mitogaku and Zen. Mitogaku refers to a school of Japanese historical and Shinto study. Modern Japanese philosophy is now also influenced by Western philosophy. Shinto is a Japanese ethnic religion focusing on ritual practices to be carried out diligently to establish a connection between present-day Japan and its ancient past. Shinto has no binding set of dogma, and the most important elements are great love and reverence for nature in all its forms, respect for tradition and family, physical cleanliness, and *matsuri* (festivals dedicated to the kami, or spirits).

Sikhism

Sikhism was founded in Punjab, India, in the late fifteenth century. Its members are known as Sikhs. The Sikhs call their faith Gurmat (Punjabi: "the way of the guru"). Sikhism was established by Guru Nanak (1469–1539 AD) and was led by a succession of nine other gurus. Sikhs believe that all ten gurus were inhabited by a single spirit. Upon the death of the tenth, Guru Gobind Singh (1666–1708 AD), the spirit of the eternal guru, transferred itself to the sacred scripture of Sikhism, Guru Granth Sahib (The Granth as the Guru), also known as the Adi Granth.

The word Sikh has its origin in the Sanskrit word *śiṣya* (disciple, student). Male Sikhs have Singh (Lion), and female Sikhs have Kaur (princess) as their middle or last name. Guru Gobind Singh commanded Sikhs to wear five *K*s (five items with the first letter *k*) all the time. They are *Kesh* (uncut hair), *Kangha* (a comb for the hair), *Kara* (an iron bracelet), *Kachera* (100 percent cotton undergarment), and *Kirpan* (an iron dagger).

Sikhism condemns caste system and believes that everyone is equal in the eyes of God. Sikhs believe in one God and discourage idol worship. By leading a virtuous and truthful life, one may end the cycle of birth and death and merge with God. The most significant religious center for the Sikhs is the Golden Temple at Amritsar in the state of Punjab in Northern India.

Islam

Islam is an Abrahamic monotheistic religion professing that there is only one incomparable God (Allah) and that Muhammad is the last messenger of God. It is the world's second-largest religion. Islam teaches that God is merciful, all-powerful, unique, and has guided humankind through prophets, revealed scriptures, and natural signs. The primary scriptures of Islam are the Koran, viewed by Muslims as the verbatim word of God, and the teachings and normative example.

Taqwa is an Islamic term for being conscious and cognizant of Allah, of truth, of the rational reality, piety, fear of God. Some descriptions of Taqwa include the following:

> God consciousness: piousness, fear of Allah, love for Allah, and self-restraint.
>
> Fear of Allah: being careful, knowing your place in the cosmos.
>
> To protect from the wrath of Allah by not indulging in things that Allah forbids.
>
> High state of heart, which keeps one conscious of Allah's presence and His knowledge.

According to the Koran, *ruh* (soul) is a command from Allah (God).

Greek Philosophy

Ancient Greek philosophy started in the sixth century BCE and continued throughout the Hellenistic period and the period in which ancient Greece was part of the Roman Empire. The Hellenistic period covers the period of ancient Greek (Hellenic) history and Mediterranean history between the death of Alexander the Great in 323 BCE and the emergence of the Roman Empire. Greek philosophy was used in a nonreligious way. It dealt with a wide variety of subjects, including political philosophy, ethics, metaphysics, ontology, logic, biology, rhetoric, and aesthetics. Greek philosophy has influenced much of Western culture.

Homer

Homer (800–900 BCE) is the name ascribed by the ancient Greeks to the semilegendary author of the two epic poems, the *Iliad* and the *Odyssey*, the central works of Greek literature. The influence of the Homeric epics on Western civilization has been incalculably vast, inspiring many famous works of literature, music, and visual art. The most important word used by Homer is *noos* (spelled "nous" in later Greek), which means "conscious mind." It comes from the word *noeein*, to see. Its proper translation in the *Iliad* would be something like perception, recognition, or field of vision.

Pythagoras

Pythagoras of Samos (570–495 BCE) was a Greek philosopher, mathematician, and the putative founder of the movement called Pythagoreanism. Pythagorean ideas exercised a marked influence on Plato and, through him, all of Western philosophy. He was one of the first to propose that the thought processes and the soul were located in the brain and not the heart. Pythagoras believed that the essence of being (and the stability of all things that create the universe) can be found in the form of numbers and that it can be encountered through the study of mathematics. For instance, he believed that things like health relied on a stable proportion of elements, with too much or too little of one thing causing an imbalance that makes a person unhealthy.

Hippocrates

Hippocrates of Kos (460–370 BCE), also known as Hippocrates II, was a Greek physician of the Age of Pericles (Classical Greece) and is considered one of the most outstanding figures in the history of medicine. The central idea of Hippocratic philosophy is the principle of wholeness, summarized by Plato in this sentence: "The certain knowledge of nature is solely possible from medicine and only when it is correctly approached as a whole." Hippocrates said authoritatively, "Some people say that the heart is the organ with which we think and that it feels pain and anxiety. But it is not so. Men ought to know that from the brain and from the brain only arise our pleasures, joys, laughter, and tears. Through it, in particular, we think, see, hear, and distinguish the ugly from the beautiful, the bad from the good, the pleasant from the unpleasant … I hold that the brain is the most powerful organ of the human body …

Eyes, ears, and tongue act in accordance with the discernment of the brain. ... To consciousness, the brain is messenger ... Wherefore I assert that the brain is the interpreter of consciousness. The diaphragm has a name due merely to chance and custom, not to reality and nature."

In the last two decades of the twentieth century, Hippocrates's conception of the brain as the messenger and interpreter of consciousness has become unbelievably complex and sophisticated. It has become a search for a neural code. Hippocrates planted the flag of consciousness on the continent of the brain, and there it remains. The precise seat of consciousness, however, continues to be elusive. Two thousand five hundred years after Hippocrates set them on their quest, the neuroscientists are still searching for consciousness.

Socrates

Socrates (469–399 BCE) was a Greek (Athenian) philosopher credited as one of the founders of Western philosophy. His doctrine is "virtue is knowledge." Socrates was occupied with the search for definitions of moral virtues. His most important contribution to Western thought is his dialectic method of inquiry, known as the Socratic method, or method of elenchus, which he largely applied to the examination of key moral concepts such as the good and justice. To solve a problem, it would be broken down into a series of questions, the answers to which gradually give the answer a person would seek.

Plato

Plato (424–347 BCE) was a philosopher in Classical Greece and the founder of the Academy of Athens, the first institution of higher learning in the Western world. He was the student of Socrates and the teacher of Aristotle. Plato expressed the idea that for something to exist, it must be capable of having an effect. So being (consciousness) is simply power. In the dialogue Sophist, written in 360 BC, Plato wrote, "My notion would be, that anything which possesses any sort of power to affect another, or to be affected by another, if only for a single moment, however trifling the cause and however slight the effect, has real existence; and I hold that the definition of being is simply power." The immortality of the soul and the idea that souls are reincarnated into different life forms are also featured in Phaedo (Plato's dialogue). Plato's conception of soul had three parts: a reason part (the part that loves truth, and rules over the other parts of the soul through the use of reason), a spirit part (the part that loves honor and victory), and an appetite part (the part that desires food, drink, and sex).

Justice will be done by the soul, in which each of these three parts does its own work and does not interfere in the workings of the other parts.

Aristotle

Aristotle (384–322 BCE), perhaps more than any other ancient Greek philosopher, discussed the problem of consciousness. Aristotle provided a first clear account of the concept of signals and information. He proposed that an event can change the state of matter and this change of state can be transmitted to other locations where the state of matter can further change. If an object that has color is placed in immediate contact with the eye, it cannot be seen. Color sets movement not in the sense organ but what is transparent—for example, the air—and that extending from the object to the organ sets the organ in movement. Sense means a power of receiving into itself the sensible forms of things without the matter.

Treatise is a formal and systematic written discourse on some subject. The treatise is near-universally abbreviated DA, for De Anima, in explaining soul-body relations. Aristotle describes that soul bears the same relation to the body that the shape of a statue bears to its material basis. Aristotle contends the following: "It is not necessary to ask whether soul and body are one, just as it is not necessary to ask whether the wax and its shape are one, nor generally whether the matter of each thing and that of which it is the matter are one. For even if one and being are spoken of in several ways, what is properly so spoken of is the actuality (De Anima ii 1, 412b6–9)." Aristotle claims, "It's clear that the soul is not separable from the body or certain parts of it, if it naturally has parts, are not separable from the body" (De Anima ii 1, 413a3–5). As a house is the form of bricks and mortars from which it is built, soul is the form of the body in much the same way. Therefore, in Aristotle's terms, the form is the actuality of the house since its presence explains why this particular quantity of matter comes to be a house as opposed to some other kind of artifact. In the same way, the presence of the soul explains why this matter is the matter of a human being, as opposed to some other kind of thing.

Aristotle describes mind (nous, often also rendered as intellect or reason) as "the part of the soul by which it knows and understands." In this way, just as having the sensory faculties is essential to being an animal, so having a mind is essential to being a human. Just as perception involves the reception of a sensible form by a suitably qualified sensory faculty, thinking involves the reception of an intelligible form by a suitably qualified intellectual faculty. Thinking involves mind becoming formed by some object of thought so that actual thinking occurs whenever some suitably prepared mind is "made like" its object by being affected by it. Thinker thinks an object only if thinker has the capacity requisite for receiving object's intelligible form. Then object acts upon that capacity by forming it. As a result, thinker's relevant capacity becomes isomorphic with that form. Aristotle maintains directly that mind is "none of the things existing in actuality before thinking." However, this has a serious problem. If the mind is indeed nothing in actuality before thinking, it is hard to understand how the hylomorphic analysis of change and affection could be explained. *Hylomorphism* is simply a compound word composed of the Greek terms for matter (*hulê*) and form or shape (*morphê*). If some dough is made in the shape of a cookie, it is actually dough before being so formed.

Regarding the relationship between soul and body, materialists hold that all mental states are also physical states. Dualists hold that the soul is a subject of mental states that can exist alone when separated from the body. Aristotle contends that it is not necessary to ask whether soul and body are one, just as it is not necessary to ask whether the wax and its shape are one. Aristotle claims, "It's clear that the soul is not separable from the body—or that certain parts of it, if it naturally has parts, are not separable from the body."

Euclid of Megara

Euclid of Megara (435–365 BCE) was a Greek Socratic philosopher who founded the Megarian school of philosophy. He was a pupil of Socrates in the late fifth century BCE. Euclid's philosophy was a synthesis of Eleatic and Socratic ideas. The Eleatics were a pre-Socratic school of philosophy founded by Parmenides (pre-Socratic Greek philosopher) in the early fifth century BC in the ancient town of Elea. Socrates claimed that the greatest knowledge was understanding the good. The Eleatics claimed that the greatest knowledge is the one universal Being of the world. Mixing these two ideas, Euclid claimed that good is the knowledge of this being. Therefore, this good is the only thing that exists and has many names but is really just one thing. Euclid adopted the Socratic idea that knowledge is virtue and that the only way to understand the never-changing world is through the study of philosophy. Euclid was also interested in concepts and dilemmas of logic.

Roman Philosophy

The most important philosophy in Rome was Stoicism, which originated in Hellenistic Greece. The Hellenistic period covers the period of Mediterranean history between the death of Alexander the Great in 323 BCE and the emergence of the Roman Empire. The contents of Stoicism represent the Roman worldview since the Stoicism insists on acceptance of all situations, including adverse ones. This seemed to reproduce what the Romans considered their crowning achievement: virtues, or manliness, or toughness. The centerpiece of Stoic philosophy was the concept of the Logos, the order God, which means rational order or meaning of the universe. Stoicism was carried to Rome in 155 AD by Diogenes of Babylon. Stoicism was perhaps the most significant philosophical school in the Roman Empire, and much of our contemporary views and popular mythologies about Romans are derived from Stoic principles. The philosophy asserts that virtue is happiness and judgment is based on behavior rather than words; that we don't control and cannot rely on external events, only ourselves and our responses.

Thomas Aquinas

Saint Thomas Aquinas (AD 1225–1274) was an Italian Catholic priest and an immensely influential philosopher and theologian. Thomas embraced several ideas put forward by Aristotle—whom he called "The Philosopher"—and attempted to synthesize Aristotelian philosophy with the principles of Christianity. Thomas believed "that for the knowledge of any truth whatsoever man needs divine help, that the intellect may be moved by God to its act." According to Thomas, the soul is not matter, not even incorporeal or spiritual matter. If it were, it would not be able to understand universals, which are immaterial. A receiver receives things that have the receiver's own nature, so for soul (receiver) to understand (receive) universals, it must have the same nature as universals. Yet any substance that understands universals may not be a matter-form composite. Therefore, humans have rational souls, which are abstract forms independent of the body. The soul exists separately from the body and after death continues in many of the capacities we think of as human.

Descartes

René Descartes (AD 1596–1650) is often credited with being the "father of modern philosophy." The first fundamental point of Descartes is the clear distinction of mind from the body. The second thing he denied is that all knowledge must come from sensation. He replaced the uncertain premises derived from sensation with the absolute certainty of the clear and distinct ideas perceived by the mind alone. Descartes established this absolute certainty in his famous reasoning: Cogito, ergo sum, or "I think, therefore I am." This became a fundamental element of Western philosophy, as it purported to form a secure foundation for knowledge in the face of radical doubt. While other knowledge could be a figment of imagination, deception, or mistake, Descartes asserted that the very act of doubting one's own existence served as proof of the reality of one's own mind; there must be a thinking entity, in this case the self, for there to be a thought.

According to Descartes, mind is by its nature not a body but an immaterial thing. Therefore, what I am is an immaterial thinking thing with the faculties of intellect and will. Mind requires nothing except God's concurrence in order to exist. But ideas are modes or ways of thinking, and therefore modes are not substances, for they must be the ideas of mind. According to Descartes, all sensation involves some sort of judgment, which is a mental mode. Hence, every sensation is, in some sense, a mental mode, and "the more attributes [that is, modes] we discover in the same thing or substance, the clearer is our knowledge of that substance." This is called the "causal adequacy principle" and is expressed as follows: "There must be at least as much reality in the efficient and total cause as in the effect of that cause," which in turn implies that something cannot come from nothing. For example, when a pot of water is heated to boil, it must have received that heat from some cause that had at least that much heat. Something that is not hot enough cannot cause water to boil because it does not have the requisite reality to bring about that effect. In other words, something cannot give what it does not have. God exists in the extended universe in a higher form, and accordingly he has the ability to cause its existence. That ultimate cause must be God, for only he has enough reality to cause it.

God cannot be the cause of human error since he did not create humans with a faculty for generating errors. Rather, humans are the cause of their own errors when they do not use their faculty of judgment correctly.

John Locke

John Locke (AD 1632–1704) was one of the most famous philosophers and political theorists of the seventeenth century. In his most important work, the Essay Concerning Human Understanding, Locke gave an analysis of the human mind and its acquisition of knowledge. According to his empiricist theory, we acquire ideas through our experience of the world. Our minds then examine, compare, and combine these ideas in numerous ways. Knowledge consists of a special kind of relationship between different ideas. Locke emphasized the philosophical examination of the human mind as a preliminary to the philosophical investigation of the world. Locke believed that before we can analyze the world, it was first necessary to examine something more fundamental: the human understanding. We have to know something about ourselves. It was "necessary to examine our own Abilities, and see, what Objects our Understandings were, or were not fitted to deal with" (Epistle 7).

The idea is the mental act of making perceptual contact with the external world object. Locke opposed the view that God planted certain ideas in our minds at the moment of creation. He made it clear that the mind has any number of inherent capacities, predispositions, and inclinations prior to receiving any ideas from sensation. Locke allowed for two distinct types of experience. Outer experience, or sensation, provides us with ideas from the traditional five senses. Inner experience, or reflection, comes from ideas such as memory, imagination, desire, doubt, judgment, and choice. Locke put forward the theory of consciousness as "internal sense" or "reflection." In that theory, consciousness is a second-order perception representing our own psychological states. Locke holds that consciousness is the employment of one's inner sense. Locke evidently thinks it is absurd, almost unintelligible, that a mental state should exist without being conscious.

Isaac Newton

Isaac Newton (AD 1642–1727) is one of the greatest scientists of the modern period, on a par with few others (perhaps Darwin or Einstein). Newton's *Principia* is the single most important work in the transformation of early modern natural philosophy into modern physical science. The four rules in *Principia* are as follows:

> Rule 1: "We are to admit no more causes of natural things than such as are both true and sufficient to explain their appearances." It means eliminate any unnecessary aspect of a theory that is not required.
>
> Rule 2: "Therefore, to the same natural effects, we must, as far as possible, assign the same causes." It means that if we observe effect B, it would have been caused by A simply because we have determined in the past that A caused B.
>
> Rule 3: "The qualities of bodies, which admit neither intensification nor remission of degrees, and which are found to belong to all bodies within the reach of our experiments, are to be esteemed the universal qualities of all bodies whatsoever." It means location is not important in physics in the sense that laws that operate on Earth will be the same that operate with the heavenly bodies, stars, and planets.
>
> Rule 4: "In experimental philosophy, we are to look upon propositions inferred by general induction from phenomena as accurately or very nearly true, notwithstanding any contrary hypothesis that may be imagined, until such time as other phenomena occur, by which they may either be made more accurate, or liable to exceptions." It simply means that any law of physics is open to future correction or improvement and Newton's laws are not to be inferred as being absolute or unconditional.

Newton saw God as the masterful Creator whose existence could not be denied in the face of the grandeur of all creation. Nevertheless, he rejected Leibniz's thesis that God would necessarily make a perfect world that requires no intervention from the Creator. However, he differed from strict adherents of deism in that he invoked God as a special physical cause to keep the planets in orbits. He warned against using the law of gravity to view the universe as a mere machine, like a great clock. He said, "This most beautiful system of the sun, planets, and comets could only proceed from the counsel and dominion of an intelligent and powerful Being." According to Newton, this Being governs all things, not as the soul of the world but as Lord over all; and on account of his dominion, he is to be called 'Lord God' or 'Universal Ruler.' The Supreme God is a Being eternal, infinite and absolutely perfect. Newton believed in the existence of spirits in the human body. He described them as of an ethereal nature and subtle enough to flow through our bodies. For him, all animal motions resulted from this spirit flowing into the motor nerves and moving the muscles by inspiration. His position is that perception and consciousness cannot possibly be explained mechanically and hence could not be a physical process.

Leibniz

Gottfried Wilhelm Leibniz (1646–1716 AD), is one of the great Renaissance men of Western thought. He has made significant contributions in several fields, including mathematics, physics, logic, ethics, and theology. Leibniz remained opposed to materialism. The realms of the mental and the physical, for Leibniz, form two distinct realms—but not in a way conducive to dualism. He opposed both materialism and dualism. His position is that perception and consciousness cannot possibly be explained mechanically and hence could not be physical processes. By means of the soul, there is a true unity that corresponds to what is called the "I" in us; such a thing could not occur in artificial machines or in the simple mass of matter, however organized it may be. Leibniz had a strong opposition to dualistic views concerning the relationship between mind and body, particularly the substance dualism that figured in the philosophy of Descartes and his followers. According to this dualism, the world fundamentally consists of two disparate substances: extended material substance (body) and unextended thinking substance (mind).

One of the better-known terms of Leibniz's philosophy, and of his philosophy of mind, is apperception, which, according to Leibniz is consciousness. He adds that this is "something not given to all souls, nor at all times to a given soul." Leibniz explained appetitions as "tendencies from one perception to another" profoundly influencing human actions. In short, we can say that in the Leibnizian realm of mind, there are indeed only perceptions and appetitions, but in these is a fundamental divide between the realm of consciousness and unconsciousness.

Voltaire

François-Marie d'Arouet (AD 1694–1778), better known as Voltaire, was a French writer, historian, and philosopher who led the eighteenth-century movement called the Enlightenment. At the center of his work was a new philosophical concept. For Voltaire, humans are natural beings governed by inexorable natural laws, and his ethics anchored right action in a self that possessed the natural light of reason inherently. Enlightenment stems from the free and public use of critical reason, and from the liberty that allows such critical debate to proceed unrestricted. Hedonistic morality was another key feature of Voltaire's Enlightenment philosophy. Hedonism is the philosophy that pleasure is the most important pursuit of mankind and the only thing that is good for an individual. Hedonists, therefore, strive to maximize their total pleasure. Hedonistic ethics was also crucial to the development of liberal political economy during the Enlightenment, and Voltaire applied his own libertinism toward this project.

Like Socrates, Voltaire was also a public critic and controversialist defining philosophy primarily in terms of its power to liberate individuals from domination at the hands of authoritarian dogmatism and irrational prejudice. Voltaire also contributed directly to the new relationship between science and philosophy that the Newtonian revolution made central to Enlightenment modernity. Especially important was his critique of metaphysics and his argument that it be eliminated from any well-ordered science.

Rousseau

Jean-Jacques Rousseau (AD 1712–1778) was a Francophone Genevan philosopher, writer, and composer. He is the writer of the "*Age of Enlightenment*." Rousseau saw a fundamental division between society and human nature and believed that man was good when in the state of nature (the condition humankind was in before the creation of civilization) but has been corrupted by the artificiality of society and the growth of social interdependence. For Rousseau, society's negative influence on men centers on its transformation of *amour de soi* (a positive self-love for self-preservation combined with the human power of reason) into *amour-propre* (artificial pride that forces a human to compare himself to others, thus creating unwarranted fear and allowing a human to take pleasure in the pain or weakness of others).

In *Discourse on the Arts and Sciences* (AD 1750), Rousseau argued that the arts and sciences had not been beneficial to humankind because they were not human needs but rather a result of pride and vanity. The opportunities they created were idleness and luxury, contributing to the corruption of man and undermining the possibility of true friendship by replacing it with jealousy, fear, and suspicion. They made governments more powerful at the expense of individual liberty. Rousseau's views on religion were highly controversial. His view that man is good by nature conflicted with the doctrine of original sin, and his theology of nature as well as the claims he made in *The Social Contract*—that true followers of Jesus would not make good citizens—led to the condemnation and banning of his books in both Calvinist Geneva and Catholic Paris.

Immanuel Kant

Immanuel Kant (AD 1724–1804) is one of the most influential philosophers in the history of Western philosophy. His contributions to metaphysics, epistemology, ethics, and aesthetics have had a profound impact on almost every philosophical movement. Kant argues that the blank slate model of the mind is insufficient to explain the beliefs about objects that we have; our beliefs must be brought by the experiences already in our minds. According to Kant, the mind itself makes its experience. It must be the mind's structuring, Kant argues, that makes experience possible. The mind brings features of experience of objects rather than given to the mind by objects. Kant gave the idea that the mind plays an active role in structuring reality, which is so familiar to us now.

For Kant, unified consciousness is a central feature of the mind. There are two kinds of consciousness: (1) consciousness of oneself and one's psychological states and (2) consciousness of oneself and one's states by acts of apperception. For Kant, this distinction between consciousness of oneself and one's states by doing acts of synthesis and consciousness of oneself and one's states as the objects of particular representations is of fundamental importance. Neither consciousness of self by doing apperceptive acts nor empirical consciousness of self as the object of particular representations yields knowledge of oneself as one is. The referential machinery used to obtain consciousness of self as subject requires no identifying (or other) ascription of properties to oneself.

According to Kant, "Most ordinary representations generated by most ordinary acts of synthesis provide the representational base of consciousness of oneself and one's states." One has the same consciousness of self no matter what else one is conscious of—thinking, perceiving, laughing, being miserable, or whatever. Kant expressed the thought this way, "When one is conscious of oneself as subject, one's bare consciousness of self, yields no knowledge of self. When we are conscious of ourselves as subject, we are conscious of ourselves as the "single common subject" of a number of representations."

According to Kant, it is impossible for us to have any experience of objects that are not in time and space. Furthermore, space and time themselves cannot be perceived directly, so they must be the form by which experience of objects can be obtained. Consciousness that apprehends objects directly, as they are in themselves and not by means of space and time, is possible, and that is by God, who has a purely intuitive consciousness. Our apprehension of objects is always mediated by the conditions of sensibility. Any discursive or concept using consciousness like ours must apprehend objects as occupying a region of space and persisting for some duration of time.

Leo Tolstoy

Count Lev Nikolayevich Tolstoy (AD 1828–1910), usually referred to in English as Leo Tolstoy, was a Russian writer who is regarded as one of the greatest authors of all time. The Tolstoyan movement is a social movement based on the philosophical and religious views of Tolstoy. Tolstoy understood religion as our "true connection" to the universe existing as God. Tolstoy wrote, "The essence of any religion lies solely in the answer to [this] question: why do I exist, and what is my relationship to the infinite universe that surrounds me? It is impossible for there to be a person with no religion [i.e., without any kind of relationship to the world] as it is for there to be a person without a heart. He may not know that he has a religion, just as a person may not know that he has a heart, but it is no more possible for a person to exist without a religion than without a heart." Tolstoy believed that science, including philosophy, cannot establish man's relationship to the infinite universe or toward its origin. Tolstoy acknowledged that the fundamental morality of all world religions is "Do unto others as would be done unto thy self."

Tolstoy believed the cause of everything is that which we call God. Neither philosophy nor science are able to establish man's relationship to the universe because this relationship must be established before any kind of philosophy or science can begin. According to Tolstoy, man lives consciously for himself but serves as an unconscious instrument for the achievement of historical, universally human goals.

Sigmund Freud

Sigmund Freud (AD 1856–1939) was an Austrian neurologist and the founder of psychoanalysis, a clinical method for treating psychopathology through dialogue between a patient and a psychoanalyst. Freud's premise is that within the human mind, there are three levels of awareness or consciousness. They are the conscious, subconscious, and unconscious. Working together, they create our reality. According to Freud, conscious mind addresses the ability to direct one's focus and the ability to imagine what is not real. According to Freud, subconscious is the storage point for any recent memories needed for quick recall. Thus the subconscious mind serves as the mind's random access memory (RAM). The unconscious mind can be seen as the source of dreams and automatic thoughts (those that appear without any apparent cause), the repository of forgotten memories (which may still be accessible to consciousness at some later time), and the implicit knowledge of things that we have learned so well that we do them without thinking. The unconscious mind is where all of our memories and past experiences reside. It's from these memories and experiences that our beliefs, habits, and behaviors are formed.

Swami Vivekananda

Swami Vivekananda (AD 1863–1902) was an Indian Hindu monk, a chief disciple of the nineteenth-century Indian mystic Ramakrishna. He was a key figure in the introduction of the Indian philosophies of Vedanta and Yoga to the Western world. According to Vivekananda, the mind is acting on three planes: the subconscious, conscious, and superconscious. Of men, the Yogi alone is superconscious. The whole theory of Yoga is to go beyond the mind. These three planes can be understood by considering the vibrations of light or sound. There are certain vibrations of light too slow to become visible; then, as they get faster, we see them as light; and then they get too fast for us to see them at all. The same holds true for sound. All knowledge that we have, either of the external or internal world, is obtained through only one method: by the concentration of the mind. No knowledge of science can be obtained unless we can concentrate our minds upon the subject. All things in nature work according to law. Nothing is excepted.

According to Vivekananda, the mind as well as everything in external nature is governed and controlled by law. Mind is the name of a stream of consciousness or thought continuously changing. No two persons have the same mind or the same body. The mind should always go toward God. According to Vivekananda, each man consists of three parts: the body, the mind, and the atman (or the self). The body is the external coating, and the mind is the internal coating of the atman, who is the real perceiver.

Bertrand Russell

Bertrand Russell (AD 1872–1970) is generally credited with being one of the founders of analytic philosophy, which holds that philosophy should apply logical techniques in order to attain conceptual clarity and that philosophy should be consistent with the success of modern science. According to Russell, it is the philosopher's job to discover a logically ideal language—a language that will exhibit the nature of the world in such a way that we will not be misled by the accidental, imprecise surface structure of natural language. Russell had great influence on modern mathematical logic. Russell held that the ultimate objective of both science and philosophy was to understand reality, not simply to make predictions. He believed in neutral monism that holds the view that the mental and the physical are two ways of organizing or describing the same elements, which are themselves neutral—that is, neither physical nor mental. This view denies that the mental and the physical are two fundamentally different things. Rather, neutral monism claims the universe consists of only one kind of stuff, in the form of neutral elements that are in themselves neither mental nor physical.

Russell said, "I ought to call myself an agnostic; but for all practical purposes, I am an atheist. I do not think the existence of the Christian God any more probable than the existence of the Gods of Olympus or Valhalla." According to Russell, sensation may be defined as the first mental effect of a physical cause and volition as the last mental cause of a physical effect.

Carl Jung

Carl Gustav Jung (AD 1875–1961) was a Swiss psychiatrist and psychoanalyst who founded analytical psychology. Jung and Freud influenced each other during the intellectually formative years of Jung's life. Jung became a proponent of Freud's psychoanalysis. He has influenced not only psychiatry but also anthropology, archaeology, literature, philosophy, and religious studies. The major concepts developed by Jung include the following:

synchronicity: an acausal principle as a basis for the apparently random simultaneous occurrence of phenomena.

archetype: a concept to denote supposedly universal and recurring mental images or themes.

archetypal images: supposedly universal symbols that can mediate opposites in the psyche.

complex: the repressed organization of images and experiences that governs perception and behavior.

extraversion and introversion: personality traits of degrees of openness or reserve contributing to psychological type.

shadow: the repressed, therefore unknown, aspects of the personality, including those often considered negative.

collective unconscious: aspects of unconsciousness experienced by all people in different cultures.

anima: in the unconscious of a man, an expression of a feminine inner personality.

animus: in the unconscious of a woman, an expression of a masculine inner personality.

self: the central overarching concept governing the individuation process, as symbolized by the union of male and female, totality, unity. Jung viewed it as the psyche's central archetype.

> individuation: the process of fulfilment of each individual which negates neither the conscious or unconscious position.

Jung regarded the psyche as made up of a number of separate but interacting systems. The three main ones were the ego, the personal unconscious, and the collective unconscious. According to Jung, the ego represents the conscious mind, as it comprises the thoughts, memories, and emotions a person is aware of. Regarding the importance of the unconscious in relation to personality, he proposed that the unconscious consists of two layers: personal unconsciousness and collective unconsciousness.

Radhakrishnan

Sarvepalli Radhakrishnan (AD 1888–1975) was an Indian philosopher and statesman and the second president of India from AD 1962 to 1967. Radhakrishnan tried to bridge Eastern and Western thought. In Radhakrishnan's philosophy, intuition, or *anubhava*—synonymously called "religious experience"—has a central place as a source of knowledge that is not mediated by conscious thought. According to Radhakrishnan, intuition is of a self-certifying character (*svatassiddha*), self-evidencing (*svāsaṃvedya*), and self-luminous (*svayam-prakāsa*). In his book *An Idealist View of Life*, he made a powerful case for the importance of intuitive thinking as opposed to purely intellectual forms of thought. Intuition plays a specific role in all kinds of experience. Radhakrishnan discerned five kinds of experience: cognitive experience, psychic experience, aesthetic experience, ethical experience, and religious experience. According to Radhakrishnan, there are five kinds of worshippers: worshippers of the Absolute; worshippers of the personal God; worshippers of incarnations like Rama, Krishna, Buddha; worshippers of ancestors, deities, and sages; and worshippers of the petty forces and spirits.

Radhakrishnan believed that there is only pure consciousness, or *caitanya*. Consciousness has no limit, and the basis of all is consciousness.

According to Mandukya, there are four states of consciousness: waking, dreaming, sleeping, and transcendental consciousness called *turya.*

Concluding Remarks

Mind, soul, and consciousness have been intriguing and puzzling topics since early civilization. Are they created by God and imparted into our body when we are born and leave the body when we die? Do they live outside the body and act on us when we are alive? Different religions, philosophies, great thinkers, and philosophers have provided their own ideas and thoughts on it. In this chapter, the ideas and views of various religions, cultures, great thinkers, and philosophers have been briefly provided. However, there is no consensus among them. Many doctrines hold the view that soul exists and is different from body, and this is called "dualism." Human beings are made up of two substances: body and soul. Soul is immortal. Most dualists agree that the soul is identical to the mind yet different from the brain or its functions. The main difficulty with dualism is the so-called "interaction problem." If the mind is an immaterial substance, how can it interact with material substances? Some dualists, however, may reply that the fact that we cannot fully explain how body and soul interact does not imply that interaction does not take place. We know many things happen in the universe, although we do not know how they happen. In the next chapter, we will further discuss monism and dualism.

CHAPTER 2

Monism and Dualism

In this chapter, we will discuss two different views of consciousness, mind, soul, and body. In order to be less verbose, we term soul, mind, and consciousness in one-word consciousness. Dualism believes that consciousness is different from the body and has no relation to the brain. Whereas monism believes that consciousness is part of the brain activity. This view is becoming even stronger by the recent progress of neurophysiology. In fact, some of them believe that the brain is a supercomputer that is capable of all our functions.

Monism

Monism is the view that a variety of existing things can be explained in terms of a single reality or substance. It is based on the concept of the monad (derived from the Greek word *monos*, meaning "single" and "without division"). The term *monism* itself is relatively recent, first used by eighteenth-century German philosopher Christian von Wolff to designate types of philosophical thought in which attempt was made to eliminate the dichotomy of body and mind. Idealistic monism holds that the mind is all that exists (i.e., the only existing substance is mental) and that the external world is either mental itself or an illusion created by the mind. Materialistic monism holds that there is but one reality, or matter, whether it be an agglomerate of atoms; a primitive world-forming substance; or the so-called cosmic nebula out of which the world evolved. It holds that only the physical is real and that the mental can be reduced to the physical. This is currently becoming the most dominant view, especially among scientists.

Scientists believe that mind, soul, or consciousness is not separate from the brain but is a part of it. They believe that consciousness comes from the chemical activity in the brain. They are trying to map the parts of brain controlling functions like thinking, memory, motor responses, sensory impressions, and so forth. Then they will artificially simulate the activity of specific neuron cells with chemicals or electrical shock to negate these neurons that affect one's feelings of anxiety, depression, and similar unwanted feelings. People could simply take a chemical in order to have a particular feeling. In this way, they can prove that consciousness or mind is not separate from the brain but is a part of it. British biologist T. H. Huxley said that all states of mind are caused by molecular changes in the brain. On this basis, the mind is just a by-product of a functioning brain. Scientists now believe that the brain is a supercomputer and consciousness or mind is a product of this supercomputer.

Dualism

Dualism, or duality, is the position that mental phenomena are, in some respects, nonphysical or that the mind and body are not identical.

Monism versus Dualism

However, can a computer have the same feeling we have when we see a sunset with red color? It can only say that it is seeing red color in the sky. Can a computer have the same feeling we have when we see the Taj Mahal? Can the computer have the same feeling when we hear Beethoven's music or Ravi Shankar's sitar? Einstein was once asked, "Do you think everything can be explained scientifically?" Einstein replied, "It would be possible to describe everything scientifically, but it would make no sense; it would be without meaning, as if you described a Beethoven symphony as a variation of wave pressure."

The experiences of enjoying something cannot be broken down into mathematical equations. But currently many scientists believe that if something cannot be broken down into mathematical equations, then it is not real—therefore, consciousness, soul, or mind does not exist. However, we are seeing that a computer or machine has no emotion, whether love or hate. Thus the scientists who are trying to prove that our responses are a mechanical reaction to sensory stimuli are simply trying to negate the idea of consciousness or soul. But they have no proof that a machine can have any feeling, whether love or hate.

Scientific View

Scientists give the opposite argument. Why is consciousness affected when changes are made to the brain if it is not part of the brain? The reason that consciousness is affected by the changes in the brain, which can be explained by a simple example of a person driving a car. Obviously, the driver is not a part of the car. If another car hits his, he will immediately say, "You hit me." It is not that the driver was hit—it was the car that was hit—but the driver identifies with the car as if he were a part of it. So the driver is affected by the changes in the car. Similarly, when the consciousness or soul depends on the body and strongly identifies with the body, it will be affected when something happens to the body, although it is actually separate from it.

According to Nobel laureate neurophysiologist Sir John Eccles, consciousness is separate from the brain. While performing experiments on the cerebral cortex, which controls movements in our body by sending signals to various muscles, he noticed that before any voluntary act is performed, the fifty million or so neurons of the supplementary motor area (SMA) within the cortex begin to act. Thus the SMA acts before the cerebral cortex, sending the necessary signals to the muscles needed to do the desired activity. Eccles therefore concluded that consciousness separate from the brain must first be there before neurological events begin. Therefore, consciousness controls brain rather than brain controlling consciousness. Philosopher Sir Karl Popper described that brain/mind or consciousness exists in two separate realities. The brain is a functioning material organ of the body, and the consciousness or mind is an immaterial entity that motivates the body.

Near-Death Experience

There are many cases of near death experiences (NDEs) with patients who were, according to all known laws of physics, technically in a state of unconsciousness or coma due to heart attack. The patients, after being brought back to consciousness, explained in detail what was happening around them while they were technically unconscious. One of the most famous proofs is the story of Maria, a migrant worker who had an NDE during a cardiac arrest at a hospital in Seattle in 1977. She later told her social worker that while doctors were resuscitating her, she found herself floating outside the hospital building and saw a tennis shoe on a third-floor window ledge, which she described in some detail. The social worker went to the window Maria had indicated and not only found the shoe but said that the way it was placed meant there was no way Maria could have seen all the details she described from inside her hospital room.

In 1991, Reynolds, then thirty-five, underwent surgery to remove a huge aneurysm at the base of her brain. Worried that the aneurysm might burst and kill her during the operation, her surgeon opted for the radical move of hypothermic cardiac arrest—chilling her body to sixty degrees Fahrenheit, stopping her heart, and draining the blood from her head. The cooling would prevent her cells from dying while deprived of oxygen. When the doctors restarted her heart and warmed her body back up, in effect she would be rebooted. To make sure that Reynolds's brain was completely inactive during the operation, the medical team put small speakers into her ears that played rapid, continuous clicks at one hundred decibels—a sound level described as equivalent to that produced by a lawn mower or a jackhammer. If any part of her mind was working, that insistent clicking would show up as electrical signals in the brain stem, which the surgeons were monitoring on an electroencephalogram. The machine confirmed that for a number of minutes, Reynolds was effectively dead in both brain and body. Yet after the surgery, she reported having had a powerful NDE, including an out-of-body experience, and accurately recalled several details about what was going on in the operating room, such as the shape of the bone saw used on her skull in addition to parts of conversations between the medical staff. Brain cells deprived of oxygen can take many hours to decay to the point of no return, especially if kept cold—hence the cases of people reviving after being buried in snowdrifts or falling into freezing lakes.

A study at the University of Michigan, published in 2013, took anesthetized rats and stopped their hearts. Within thirty seconds, the rats' EEG brain signals flatlined, but first they spiked, with an intensity that suggested that different parts of the brain were communicating with one another even more actively than when the rats were awake.

If we could establish that spikes in neural activity occur in a dying human brain like the ones George Mashour, an anesthesiologist from the University of Michigan, and his colleagues saw in rats, that could both help explain near-death experiences and give us some clues about the neurobiological nature of consciousness. Due to obvious ethical reasons, it is not possible to carry out similar experiments on humans. Mashour says it's unlikely that we can collect enough useful data on people who've had NDEs in the midst of a cardiac arrest and lived to tell the tale. But his study on rats, he says, at least "illuminated the possibility" that to explain near-death experiences you need not "abandon the connection between the brain and consciousness."

Subjective and Objective View of Soul

In the absolute sense, the soul can only be conscious of itself because it alone exists as that state of pure singularity. When we say it is "conscious of itself," we separate the intellectual level into two aspects: the aspect as observer and the aspect as observed. Intellectual examination, in fact, reveals the existence within consciousness of three values inherent in any process of conscious experience or any process of observation: (1) the observer, (2) the observed, and (3) the process of linking the observer and the observed.

Even though there is but one consciousness, this principle of three values emerges. Consciousness being awake to itself experiences itself and is the knower, the process of knowing, and the known; observer, process of observation, and observed; or subject, object, and the process of linking them. In this state of absolute consciousness, these three values are the same, yet they represent these aspects of the same singularity.

It is obvious that every relative experience requires a subject coming together with an object. This coming together takes place both on the level of attention as well as on the sensory level of perception. When the subject comes together with the object through the process of observation, then the experience occurs and the knowledge of the object by the subject takes place. Therefore, knowledge is the result of the coming together of the observer, the process of observation, and the observed.

As one consciousness leads to three aspects, the interaction between the three and the resultant aspects, relationships, and their interaction, and so forth, leads to an infinite number of ever-expanding possibilities. All these possibilities, all these forces of interaction and relation, exist in the soul. The interaction of forces, even though within the soul, creates a dissymmetry, as if a distortion, in the flat and homogeneous—yet infinitely flexible—absolute singularity of soul. The virtual pull and push, rise and fall, vibration and silence, and dynamism and silence lead to the formation of structure within the soul.

Hinduism View

Hindu philosophy has two separate views: Dvaita and Advaita. Dvaita is dualism, meaning God and the individual soul and matter are two separate entities. Vishnu or Brahman is the supreme self, the absolute truth. Advaita is monism with the belief that God and soul are one. The world is an illusion. When the soul releases itself from illusion, it merges itself with God.

Important views of consciousness in the Hindu tradition is found in the Samkhya and Yoga schools of Indian philosophy. This school is believed to have been founded by the philosopher Kapila, who may have lived in the seventh century BCE. The Samkhya tradition includes six classical schools of Hindu philosophy: Samkhya, Yoga, Nyaya, Vaisheshika, Mimamsa, and Vedanta.

Samkhya is based on a dualistic metaphysics approach composed of two irreducible, innate, and independent realities: (1) consciousness or spiritual self (atman or purusha) and (2) unconscious primordial materiality (prakriti or *Pradhana*). Nature is the universal material substratum out of which all phenomena other than consciousness emerges. These phenomena are physical transformations of three qualities (guna): (1) harmony, purity, and virtue (sattva); (2) energy and activity (raja); and (3) inertia and obstruction (tamas). The unconscious primordial materiality, Prakriti, contains twenty-three components, including intellect (buddhi, *mahat*), ego (*ahamkara*), and mind (manas). Therefore, the intellect, mind, and ego are all forms of unconscious matter. Illumination from purusha makes thought processes and mental events conscious. Hence, consciousness is similar to light, which illuminates the material configurations, or "shapes," assumed by the mind. Therefore, intellect, after receiving cognitive structures, form the mind and illumination from pure consciousness, creating thought structures that appear to be conscious. Ahamkara, the ego or the phenomenal self, appropriates all mental experiences to itself and thus personalizes the objective activities of mind and intellect. But consciousness is itself independent of the thought structures it illuminates. By including mind in the realm of matter, Samkhya-Yoga avoids Cartesian dualism.

Hindu philosophy does not believe that consciousness or mind is created by the neural activities of brain. Mind becomes conscious by borrowing consciousness from the only source, which is Brahman, the supreme self. Brahman, or supreme consciousness, is present in every being and thing. According to Hindu Upanishads, there are four states of consciousness: waking (conscious), dreaming (unconscious), deep sleep (subconscious), and turiya (pure consciousness). Turiya transcends all the other three states of consciousness to attain pure consciousness.

Buddhism View

Buddhism is based on the opposite view of Hinduism—that is, no separate entity for consciousness or atman. It is based on anatman. According to Buddhism, thoughts themselves are the thinker, and the experiences are the experiencer. Abhidharma is one of the oldest Buddhist philosophies and can be traced back to the third century BCE. When Abhidharma analyzes a matter, it takes a matter as not as composition of stable particles but as fleeting material events instantaneously coming into and going out of existence depending on causes and conditions. Similarly, the mind is analyzed into basic types of states or events that make up the complex phenomenon called mind. These states can be observed by turning inwardly and attending to the way we feel, remember, think, and so forth. By doing this, we notice a variety of states of awareness changing rapidly. Abhidharma identifies these changing mental states as the basic elements of mind. According to Abhidharma, each mental state has two aspects: (1) primary factor of awareness (*citta*), whose function is to be aware of the object, and (2) mental factors (*caitesika*), whose function is to qualify the awareness by determining the characteristic and quality of the object. Awareness, or citta, is also described as *vijnana*, which is translated as consciousness or awareness. In Abhidharma, there are six types of awareness: five from the physical senses (sight, hearing, sense, smell, and touch), with the sixth being mental cognition or awareness.

The Buddhist tradition conceives the human individual as consisting of five types of aggregates that serve as the basis of designating a person: (1) material form or body (*rūpa*), (2) sensations (*vedanā*), (3) apperception (*saṃjña*), (4) volitions or dispositional formations (*saṃskāra*), and (5) consciousness (*vijñāna*). Form includes the sensory systems, which form an anatomical and physiological point of view of matters. Sensations define the quality of the impressions that result from contact between the sense and its object. Apperception refers to the capacity to comprehend the specific marks of phenomenal objects. Volitions are primarily responsible for bringing forth future states of existence. They include all the conditioned factors that are intrinsic to consciousness as well as factors that are dissociated from consciousness.

Lastly, consciousness is defined as the impression of each object or as the bare apprehension of each object. Unlike sensation and apperception, which apprehend the specific characteristics of objects, consciousness acts as an integrating and discerning factor of experience.

Views of other philosophers

Plato believed that the true substances are not physical bodies, which are ephemeral, but the eternal forms of which bodies are imperfect copies. These forms not only make the world possible; they also make it intelligible, because they perform the role of universals. Their connection with intelligibility is relevant to the philosophy of mind. Because forms are the grounds of intelligibility, they are what the intellect must grasp in the process of understanding. Plato presents a variety of arguments for the immortality of the soul, but the one that is relevant for our purposes is that the intellect is immaterial because forms are immaterial and intellect must have an affinity with the forms it apprehends. This affinity is so strong that the soul strives to leave the body in which it is imprisoned and to dwell in the realm of forms. It may take many reincarnations before this is achieved. Therefore, Plato's dualism is not simply a doctrine in the philosophy of mind, but an integral part of his whole metaphysics. One of the most famous philosophers who believed in dualism and who is often regarded as the father of this mind-body problem is a French philosopher named René Descartes.

The problem of consciousness: what is consciousness? How is it related to the brain and the body? The problem of intentionality: what is intentionality? How is it related to the brain and the body? The problem of the self: what is the self? How is it related to the brain and the body? The problem of embodiment: what is it for the mind to be housed in a body? What is it for a body to belong to a particular subject? The seemingly intractable nature of these problems has given rise to many different philosophical views.

From the mid-seventeenth century through the late nineteenth century, consciousness was widely regarded as essential or definitive of the mind. René Descartes defined the very notion of thought (pensée) in terms of reflexive consciousness or self-awareness. In the *Principles of Philosophy* (1640), he wrote, "By the word 'thought' (pensée), I understand all that of which we are conscious as operating in us." John Locke offered a similar if slightly more qualified claim in *An Essay on Human Understanding* (1688): "I do not say there is no soul in man because he is not sensible of it in his sleep. But I do say he cannot think at any time, waking or sleeping, without being sensible of it. Our being sensible of it is not necessary to anything but our thoughts, and to them it is and to them it always will be necessary."

Leibniz offered a theory of mind in the *Discourse on Metaphysics* (1686) that allowed for infinitely many degrees of consciousness and perhaps even for some thoughts that were unconscious, the so-called "petites perceptions." Leibniz was the first to distinguish explicitly between perception and apperception—that is, roughly between awareness and self-awareness. Immanuel Kant (1787) argued that an adequate account of experience and phenomenal consciousness required a far richer structure of mental and intentional organization. Phenomenal consciousness, according to Kant, could not be a mere succession of associated ideas but at a minimum had to be the experience of a conscious self situated in an objective world structured with respect to space, time, and causality.

At the outset of modern scientific psychology in the mid-nineteenth century, the mind was still largely equated with consciousness. T. H. Huxley remarked, "How it is that anything so remarkable as a state of consciousness comes about as a result of irritating nervous tissue is just as unaccountable as the appearance of the djin when Aladdin rubbed his lamp."

Consciousness has three relevant questions:

> The descriptive question: What is consciousness? What are its principal features? And by what means can they be best discovered, described, and modeled?
>
> The explanatory question: How does consciousness of the relevant sort come to exist? Is it a primitive aspect

of reality, and if not, how does (or could) consciousness in the relevant respect arise from or be caused by nonconscious entities or processes?

The functional question: Why does consciousness of the relevant sort exist? Does it have a function, and if so, what is it? Does it act causally, and if so, with what sorts of effects? Does it make a difference to the operation of systems in which it is present, and if so, why and how?

The three questions focus respectively on describing the features of consciousness, explaining its underlying basis or cause, and explicating its role or value.

So far, it looks unlikely that any single theoretical perspective suffices for explaining all the features of consciousness that we wish to understand.

Modes of Knowledge

Physicists found two modes of knowing—one is symbolic or inferential; the other is intimate or direct. In Christianity, symbolic knowledge has been called "twilight knowledge," in which creation is perceived by clearly distinguished ideas. The second is called "daybreak knowledge," in which all ideas are rejected, only considering that God is one.

Taoism recognizes two general forms of knowing: conventional knowledge and natural knowledge. For us, all knowledge according to Taoism is conventional knowledge because we do not feel that we really know anything unless we can represent it to ourselves in words, mathematics, or music. The natural knowledge is aimed at understanding of knowledge directly instead of representing it in some forms.

These two forms of knowledge are also clearly distinguished in Hinduism, as stated in Mundaka Upanishad, as lower mode and higher mode. The lower mode, termed *aparavidya*, corresponds to symbolic knowledge. It is conceptual and comparative knowledge and is based on the distinction between knower (*pramatr*) and the known (*visaya*).

The higher mode, termed *paravidya,* is reached all at once, intuitively and immediately, and not by progressive movement through the lower mode aparavidya.

In Mahayana Buddhism, the symbolic mode and intimate mode are termed vijnana and prajna, respectively. Academic and environmental activist David Sujuki elaborates: "Prajna goes beyond vijnana. We make use of vijnana in our world of senses and intellect, which is characterized by dualism in the sense that there is one who sees and there is the other that is seen—the two standing in opposition. In prajna, this differentiation does not take place. What is seen and the one who sees are identical; the seer is the seen and the seen is the seer.

Many consciousness researchers have argued that any action of a nonphysical mind on the brain would violate physical laws, such as the conservation of energy. Their argument is that any mental decision to do something would cause something totally nonphysical and would cause a group of neurons to fire. This goes against the fundamental law of physics. But neuroscience has progressed tremendously. We now understand the different parts of the brain, its structures and functions. We have gone a long way in understanding the functions and firing patterns of a hundred billion neurons in the brain, their one thousand trillion connections between each other through axons and dendrites. Decision-making tasks are determined by the neural activities within the prefrontal cortex of the brain. An average adult brain weighs about three pounds (thirteen to fourteen hundred grams)—about 2 percent of the total body weight—but consumes up to 20 percent of body energy. Hence, consciousness created by the brain does not violate the fundamental law of physics. The function of the brain and its relation to consciousness has been discussed in detail in chapter 8.

Occam's razor cannot consistently be put forward by a physicalist or materialist as a justification of the belief that dualism is false. Occam's razor, attributed to English philosopher William of Ockham (AD 1287–1347), is a principle that states that supposing two explanations for an occurrence exist, then the simpler one is usually better. If one applies Occam's razor unrestrictedly, then it recommends monism until dualism receives either more support or proof.

According to the theory of evolution, which is the most likely scientific explanation of the origin of life according to current evidence, man evolved from simpler organisms, which in turn evolved from even simpler one-celled organisms, which in turn evolved from nonliving proteins through some complex but entirely physical process. Now, if nonliving proteins are purely physical and have no dualistic mind, then how can the physical process of evolution instill this nonphysical soul into our ancestry somewhere along the line? And if it did, at what point did it do so? A similar argument can be made from the development of a new human being. People start out as the fusing of two cells. The cells combine genetic structure and then divide repeatedly through wholly physical means. A fetus develops, then it is born, and it grows from a baby into an adult. Now, if the adult has this soul, at what point did he acquire it? The reasoning and implications are the same as above. Dualism relies on some unknowable intervention in the creation or addition of a soul or mind to man; monism needs no divine influence to describe the actions of man.

Neither dualism nor monism are complete or accurate in their description of human performance. Moreover, both fail at being truly scientific. Dualism is not scientific in any sense of the word; it denies the validity of observation, what all science is built upon, and puts forth a mere theory, the nonphysical soul, with no evidence other than reasoning. Monism is also suspect. Just because the soul can't be proven from current observation does not mean that it's not true, and Ockham's razor is therefore not a proof of monism but merely an argument for it. Monism is too quick to deny something it cannot prove or disprove.

The criticisms of both dualism and monism are too strong to take either one as truth. Instead, a better philosophy would combine the methods of both dualism and materialism but leave out their shortsightedness. Instead of denying the validity of anything not absolutely known, it would assign a degree of trust to it, depending on how likely it is true. It would use monism's empirical approach but not deny anything it cannot prove or disprove and never assert anything to be absolutely proven, instead being proven to a high degree of probability.

Concluding Remarks

The most contentious issue about the consciousness is dualism versus monism. All early philosophies believe in some kind of dualism. According to these philosophies, the brain does not generate consciousness or mind. Monism is believed by the neurophysiology and associated science by the recent progress in neurophysiology and computer science. They cling stubbornly to mechanistic and physical explanations of life and reject any nonmechanistic idea. But consciousness, or mind, is real. It has been seen in numerous cases by physicians in numerous cases that when patients are medically unconscious, even then they can accurately describe what was happening when they were unconscious.

In the next chapter, I will describe artificial intelligence and consciousness. Currently, artificial intelligence is a very active research area of scientists.

CHAPTER 3

Consciousness and Artificial Intelligence

Artificial intelligence (AI) is intelligence exhibited by machines. AI is a field of computer science that explores models of complex problem-solving by computers. The overall research goal of artificial intelligence is to create technology that allows computers and machines to function in an intelligent manner. The general problem of simulating (or creating) intelligence has been broken down into subproblems. These consist of particular traits or capabilities that researchers expect an intelligent system to display. The traits are as follows: reasoning, problem-solving, knowledge, reasoning, planning, learning, language processing, perception, social intelligence, and creativity. Rapid progress in computer technology has created intense research in artificial intelligence. Few scientists think that computers cannot only possess consciousness but also far exceed human intelligence.

History of the Computer

The Turing machine, developed by Alan Turing in the 1930s, is a theoretical device that consists of tape of unlimited length that is divided into little squares. Each square can either hold a symbol (1 or 0) or be left blank. A read/write device reads these symbols and blanks, which gives the machine its instructions to perform a certain program. Nowadays, people in developed countries have access to computers.

As we all know, a computer is a device that can be instructed to carry out an arbitrary sequence of arithmetic or logical operations called program automatically. This makes computers applicable to a wide range of tasks, and they have become essential in our lives. English mathematics professor Charles Babbage invented the first mechanical computer, and it is called the father of the computer. His mechanical computer led to the complex electronic computer. The first electronic digital computer was built at Iowa State University by John Vincent Atanasoff and Clifford Berry. The computer was built from vacuum tubes, rotating drums for memory, and recording numbers by scorching marks on cards. It weighed 750 pounds. After transistors were invented, the first computer using transistors was introduced in the market in 1951. Tapes and disks were used as storage media. With the invention of integrated circuits, computers became much more powerful, smaller, faster, and more reliable. Microsoft Disc Operating System (MS-DOS) came along in 1980. IBM introduced the personal computer (PC) for home and office use in 1981. In 1984, Apple introduced the Macintosh computer with its icon- driven interface. In 1985, the first Windows operating system was launched.

Computer Technology

The main structural elements of a computer as shown in figure 3 are as follows:

central processing unit (CPU)—data processing and control

main memory (primary storage)—stores data

secondary storage—stores permanent data

input and output (I/O) devices—moves data between the computer and its external environment

external devices—printer, computer monitor for display

system interconnection—provides mechanism for communication among different parts

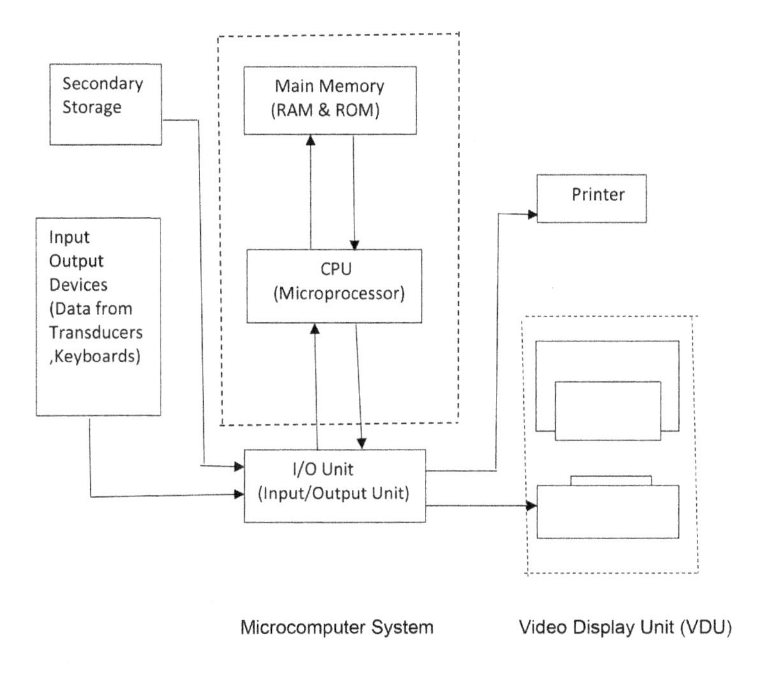

Figure 3: Structure of a computer

The two components at the heart of the hardware structure are the central processing unit (CPU) and the main memory. The CPU monitors and controls the operation of the other devices and the flow of information to and from these devices, and it performs all the necessary manipulations of the data. The CPU uses the main memory to store information for immediate access. Secondary storage devices provide permanent storage of large amounts of data. Secondary storage is also called secondary memory, external memory, backing store, or auxiliary storage. This storage may consist of magnetic tapes, a magnetic disk, an optical memory device, or a similar device. Input/output devices provide an interface between the computer and the user. There is at least one input device (keyboard, mouse) and at least one output device (printer, display screen). Input and output devices like keyboards and printers, together with the external storage devices, are referred to as peripherals. The computer system requires interconnections between the various components. When these data paths carry more than one bit simultaneously from a number of different components, it is referred to as a data bus.

Binary bit consists of binary 1, which corresponds to logical high, and binary 0, which corresponds to logical low. Qubit is the short form of quantum bit used in the quantum computer. In qubit format, 1 "qubit," or quantum bit, can represent both 0 and 1 simultaneously—that is, superposition of both. We will briefly describe the quantum computer later.

An integrated circuit or monolithic integrated circuit (also referred to as an IC, a chip, or a microchip) is a set of electronic circuits on one small flat piece (or "chip") of semiconductor material, normally silicon. The integration of large numbers of tiny transistors into a small chip resulted in circuits that were orders of magnitude, smaller, cheaper, and faster than those constructed of discrete electronic components.

Computers are getting more and more powerful with advances in integrated circuit technology. Moore's law is the observation that the number of transistors in a dense integrated circuit doubles approximately every two years. The observation is named after Gordon Moore, the cofounder of Fairchild Semiconductor and Intel. When my team and I designed an IC with one million transistors in 1998, we thought it was a great achievement. But now, as of 2016, the largest transistor count in a commercially available single-chip processor is over 7.2 billion—the Intel Broadwell-EP Xeon. In other types of ICs, such as field-programmable gate arrays (FPGAs), Intel's (previously Altera) Stratix 10 has the largest transistor count, containing over thirty billion transistors. During Intel's CES 2014 keynote, Senior Vice President Mooly Eden said it would take only twelve years for the company to make a processor that features as many transistors as there are neurons in a human brain—one hundred billion neurons.

Turing Machine

A Turing machine is an abstract mathematical model of computation invented in 1936 by mathematician Alan Turing. An abstract machine, it manipulates symbols on a strip of tape according to a table of rules. The tape divides into cells, one next to the other. Each cell contains a symbol from some finite alphabet. The alphabet contains a blank symbol (here written as 0) or any other symbol. The tape is assumed to be arbitrarily extendable to the left and to the right—the Turing machine is always supplied with as much tape as it needs for its computation. Cells that have not been written before are assumed to be filled with the blank symbol. Figure 4 gives hypothetical representation of the machine.

Figure 4: Turing tape

In some models, the tape has a left end marked with a special symbol; the tape extends or is indefinitely extensible to the right. The machine has a head that can read and write symbols on the tape and move the tape left and right one (and only one) cell at a time. In some models, the head moves and the tape is stationary. The machine has a state register that stores the data. This is a finite table of instructions that, given the state the machine is currently in and the symbol it is reading on the tape (symbol currently under the head), tells the machine to do the following in sequence: Either erase or write a symbol and then move the head. Then assume the same or a new state as prescribed.

Turing machine's operations are "syntactical," meaning they only recognize symbols and not the meaning of those symbols—that is, their semantics. Even the word "recognize" may be misleading, as it implies a subjective experience. So the better word would be sensitive to the symbols, whereas the brain is capable of semantic understanding.

Quantum Computer

Today's computers are digital. It is binary working on two bits, 0 and 1—that is, a bit could be either 0 or 1. Normally we count in base 10, called decimal or denary count. We count from 0 to 9, then back to 1, adding a 0 after that for 10, and so on. In binary count, we count 0, then 1, then 10, then 11, and so on. Hence the computers are limited to two states. Quantum computers are not limited to two states. They use qubits instead of binary bits. Qubit is a superposition of both 0 and 1. To visualize qubit, let us think of think of an electron that could either be in spin-up or spin-down state. Changing the electron from spin-up to spin-down state would need a pulse of energy. However, if the pulse of energy is not strong enough to change the state of the electron, then it will behave as if it is in both states of 0 and 1—that is, superposition of 0 and 1. Hence, it gives the quantum computer enormous processing power. A 30-bit quantum computer could do 230 bits or 10 terabits, processing simultaneously.

Another aspect of quantum computer is entanglement. When two electrons, one spin up and the other spin down, are moved far apart, then measuring one electron will instantaneously affect the other electron, irrespective of the distance between them. This is called quantum entanglement. This gives the quantum computer enormous speed. In normal computers, current has to move along the track to reach another part of the computer chip. In quantum computers, the speed is unlimited. Taken together, qubit and quantum entanglement give an enormous processing power for the quantum computer.

Quantum computers could one day replace silicon chips, just as the transistor once replaced the vacuum tube. In 2007, Canadian start-up company D-Wave demonstrated a sixteen-qubit quantum computer. The computer solved a sudoku puzzle and other pattern-matching problems. In February 2012, IBM scientists said that they had made several breakthroughs in quantum computing with superconducting integrated circuits. In October 2015, researchers at University of New South Wales built a quantum logic gate in silicon for the first time. In August 2016, scientists at the University of Maryland successfully built the first reprogrammable quantum computer. Quantum computers must have at least several dozen qubits to be able to solve real-world problems and thus serve as a viable computing method. Therefore, we are still few years away from having a viable quantum computer.

Consciousness in Artificial Intelligence

Artificial intelligence is intelligence demonstrated by a machine usually having a computer. This has become a hot subject of research nowadays, some thinking it could exceed human intelligence and cause a potential danger to the human race. There is no doubt that with the speed and processing power of current computers and potential quantum computers, machines with artificial intelligence can perform much faster than human beings can. However, can artificial intelligence invent new things, create new ideas, new theories, new philosophies, and so on? Can artificial intelligence propose theories like general theory of relativity, quantum mechanics, Bohr's atomic model, and so forth? We can make artificial intelligence powerful and extremely fast, but it is still a stupid machine according to famous scientist Feynman.

In humans and animals, we know that the specific content of any conscious experience—the deep blue of an alpine sky or the fragrance of jasmine in the night air—is furnished by parts of the cerebral cortex associated with thought, action, and other higher brain functions. If a sector of the cortex is destroyed by stroke or some other calamity, the person will no longer be conscious of whatever aspect of the world that part of the brain represents. For instance, a person whose visual cortex is partially damaged may be unable to recognize faces, even though he can still see eyes, mouths, ears, and other discrete facial features. Consciousness can be lost entirely if injuries permanently damage most of the cerebral cortex. Lesions of the cortical white matter, containing the fibers through which parts of the brain communicate, also cause unconsciousness. In addition, small lesions deep within the brain along the midline of the thalamus and the midbrain can inactivate the cerebral cortex and indirectly lead to a coma—and a lack of consciousness.

Consciousness also requires the cortex and thalamus to be constantly suffused in a bath of substances known as neuromodulators, which aid or inhibit the transmission of nerve impulses. (Readers are advised to read section 8.1 to familiarize with the structure of brain.) Also, whatever the mechanisms necessary for consciousness, we know they must exist in both cortical hemispheres independently. Widespread damage to the cerebellum, the small structure at the base of the brain, has no effect on consciousness, despite the fact that more neurons reside there than in any other part of the brain.

Consciousness is subjective. It depends on each individual. One person may be excited by seeing his child or wife, whereas another person who has no relation with him or her will ignore it. The function of our brains creates consciousness. A recent study shows that consciousness involves cortex areas of the brain, including the frontal lobe, parietal lobe, occipital lobe, and the thalamus.

A person becomes knowledgeable about all aspects of the color red, the type that computers can learn—wavelengths, frequencies, and so forth. However, the person is not learning what red looks like in reality, for she has not seen it. Therefore, she does not know what red truly looks like. No amount of factual data can give her or a computer a subjective experience of red. This subjective experience can only come from consciousness, which computers cannot have. Artificial intelligence cannot have consciousness.

Philosopher David Chalmers has posed "the hard problem of consciousness," asking why all of this information processing needs to feel a certain way to us from the inside. The problem of AI consciousness is not just Chalmers's hard problem applied to the case of AI, for the hard problem of consciousness assumes that we are conscious. After all, each of us can tell from introspection that we are now conscious. It asks why we are conscious. Why does all our information processing feel a certain way from the inside? In contrast, the problem of AI consciousness asks whether AI, being silicon based, is even capable of consciousness. It does not presuppose that AI is conscious—that is the question. These are different problems, but they are both problems that science alone cannot answer.

Consciousness may be limited to carbon substrates only. Carbon molecules form stronger, more stable chemical bonds than silicon, which allows carbon to form an extraordinary number of compounds, and unlike silicon, carbon has the capacity to form double bonds more easily. This difference has important implications in the field of astrobiology, because it is for this reason that carbon, and not silicon, is said to be well suited for the development of life throughout the universe. If the chemical differences between carbon and silicon affect life itself, we should not rule out the possibility that these chemical differences also impact whether silicon gives rise to consciousness, even if they do not hinder silicon's ability to process information in a superior manner.

Concluding Remarks

In this chapter, I have provided a brief description of the history and structure of computer and a brief introduction of quantum computer that is still in primary state. Then I have given the current research work of artificial intelligence and ongoing logistics of consciousness in artificial intelligence. The open question is whether nonbiological machines can support consciousness since these have to duplicate the essential electrochemical processes that are occurring in the brain during consciousness. If this were possible at all without organic materials, then it would need more than Turing machines or quantum computers because they are syntactic processors (symbol manipulators).

CHAPTER 4

Theories of Consciousness

Consciousness can be described as taking unconnected and chaotic properties or bits of data and correlating them into an ordered structure or schema. Consciousness is thus antientropic because it makes an unordered structure become ordered—that is, entropy reversed. Hence, consciousness is some kind of energy.

Consciousness was not a part of normal science to carry out any objective experimentation. Scientists looked at it as a hobby for some philosophers. However, it all changed in the late 1980s. Today consciousness is at the forefront of neuroscience research, with exciting challenges. The focus is how our brains create consciousness. Consciousness is subjective; it is mainly related to the sense of self. At any given time, a massive flow of sensory information reaches our senses. But our consciousness deals with only a small amount of it, one item at a time. We are fundamentally limited to one conscious thought at a time. Attention serves as the gateway of consciousness.

In this chapter, we will briefly discuss some of the theories proposed to solve the mystery of consciousness before neuroscience started seriously researching the riddle of consciousness.

Integrated Information Theory

According to integrated information theory (IIT), consciousness is an intrinsic, fundamental property of any physical system. IIT was proposed by Giulio Tononi. IIT accepts the existence of consciousness as certain and then tries to find out what kinds of properties the subject would require to account for it. Hence, it attempts to identify the essential properties of conscious experience and, from there, the essential properties of conscious physical systems. The essential properties of experience are axioms. The axioms are as follows.

> Intrinsic existence: Consciousness exists; each experience is actual and independent of external observers.
>
> Composition: Consciousness is structured. Each experience is composed of multiple phenomenological distinctions.
>
> Information: Consciousness is specific. Each experience is composed of a specific set of specific phenomenal distinctions, thereby differing from other possible experiences. For example, an experience may include phenomenal distinctions specifying a large number of spatial locations, such as a bedroom (as opposed to no bedroom), a bed (as opposed to no bed), a book (as opposed to no book), as well as many negative concepts, such as no bird (as opposed to a bird), no bicycle (as opposed to a bicycle), no bush (as opposed to a bush), and so on. In that way, it necessarily differs from a large number of alternative experiences.
>
> Integration: Consciousness is unified: each experience cannot be reduced to noninterdependent, disjoint subsets of phenomenal distinctions. Thus I experience a whole visual scene, not the left side of the visual field or the right side independent of each other.

> Exclusion: Consciousness is definite in content, space, and time. Each experience has the set of phenomenal distinctions. It is neither a subset nor a superset.

The above axioms capture the essential properties of every experience. But for each experience, there is a causal property of physical substrate called postulate. The postulates are as follows:

> Intrinsic existence: The system must be actual, with cause-effect power.
>
> Composition: The system must be structured composed of subsets in various combinations having cause-effect power within the system.
>
> Information: The system must specify a cause-effect structure in a particular way, differing from other possible ones (differentiation).
>
> Integration: The cause-effect structure specified by the system must be unified.
>
> Exclusion: The cause-effect structure specified by the system must be definite, specified over a single set of elements, neither less nor more.

Integrated information theory has received both broad criticism and support. Neuroscientist Christof Koch, who has helped to develop the theory, has called IIT "the only really promising fundamental theory of consciousness. Some critics have challenged that IIT proposes conditions that are necessary for consciousness but are not sufficient. Objections have also been made to the claim that all of IIT's axioms are self-evident. Disagreements over the definition of consciousness also lead to inevitable criticism of the theory.

Global Workspace Theory

Global workspace theory (GWT) is a simple cognitive architecture that has been developed to account qualitatively for a large set of matched pairs of conscious and unconscious processes. Bernard Baars proposed it. Consciousness is associated with a global broadcasting system that disseminates information through the brain. The easiest way to think about GWT is in terms of a "theater metaphor." In the "theater of consciousness," a "spotlight of selective attention" shines a bright spot on stage. The bright spot reveals the contents of consciousness—actors moving in and out, making speeches, or interacting with each other. The audience is not lit up, is in the dark (i.e., unconscious), and is watching the play. Behind the scenes, also in the dark, are the director (executive processes), stagehands, scriptwriters, scene designers, and the like. They shape the visible activities in the bright spot but are invisible themselves. GWT contents are proposed to correspond to what we are conscious of and are broadcast to a multitude of unconscious cognitive brain processes, which may be called receiving processes. Other unconscious processes, operating in parallel with limited communication between them, can form coalitions that can act as input processes to the global workspace.

GWT theory relies on three theoretical constructs: unconscious specialized processors, a conscious global workspace, and unconscious contexts. Figure 5 shows the major constructs in GWT theory and the functional relations between them. GWT relies on theoretical constructs: unconscious specialized processor, global workspace, and context.

Unconscious Specialized Processor

These experts in the brain could be single cells such as cortical feature detectors for colors, faces, and line orientation. In addition, they could be an entire network of neurons such as cortical columns or the Broca region in the frontal lobe of the dominant hemisphere (usually the left) of the hominid brain, with functions linked to speech production. They could also be Wernicke's area, located in the temporal lobe on the left side of the brain, responsible for the comprehension of speech and basal ganglia situated at the base of the forebrain and strongly connected with the cerebral cortex, thalamus, and other brain areas. By posting messages in the global workspace, they can recruit a coalition of other experts by sending messages.

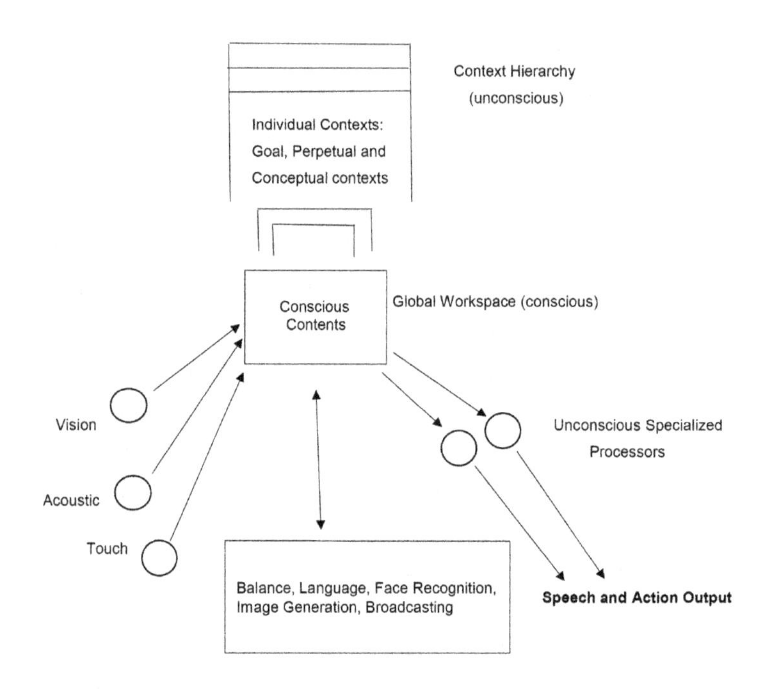

Figure 5: Global workspace architecture

It is an architectural capability for system-wide integration and dissemination of information, as shown in figure 5. Since conscious experience seems to be focused primarily toward perception, it is convenient to assume that preperceptual processors—visual, auditory, and multimodal—compete for access to global workspace.

Context: These expert processors provide the directors, playwrights, and stagehands behind the scenes of the theater of mind. They are structures with conscious contents without being conscious themselves, just as the playwright determines the words and actions of the actors on stage without being visible. Contexts may be momentary—just the first word in a sentence sets the interpretation for the second word—or they could be long lasting just like love expectations, professional expectations, worldviews, and other things people care about.

The conscious events can set up unconscious contexts. Consciousness from years ago can influence one's current experience, even if the memories of the earlier events do not become conscious again. For example, a shocking or traumatic event in early life can set up largely unconscious expectations that may shape subsequent conscious experiences.

McGinn's Theory of Consciousness

Colin McGinn's theory has two central tenets. First, the consciousness phenomenon is perfectly natural and in no way mysterious. Second, we are incapable of understanding the riddle of the consciousness. McGinn's theory has the concept of cognitive closure, which is lack of the procedure for concept formation. McGinn admits that the brain produces consciousness. However, he claims that we cannot understand how it does it. His argument is that we do not have a single mechanism that can access both brain and consciousness. Our access to consciousness is through the faculty of introspection. Our access to the brain is through using our senses. However, senses do not give us access to the consciousness, and introspection does not give us access to the brain. Thus McGinn's argument is follows:

> We have introspective access to consciousness but not to the brain.
>
> We have extrospective access to the brain but not to consciousness.
>
> We have no accessing method that is both introspective and extrospective.
>
> We have no method that can give us access to both consciousness and brain.

McGinn assumes that consciousness states are caused by brain states, but it rejects that consciousness states are identical to the brain states. He presupposes the view that consciousness cannot be simply identified with the physical. Thus McGinn is committed to dualism. There are two different states a person has: brain states and consciousness states.

David Chalmers's Theory of Consciousness

David Chalmers proposes consciousness as a fundamental building block like space, time, mass, charge. These building blocks cannot be explained by anything more basic. There are fundamental laws like laws of gravity, laws of motion, laws of electromagnetic waves, and so forth. These are essential properties of the universe and cannot be explained by anything more fundamental. The list of fundamentals will expand as science advances. Consciousness is also a fundamental building block of nature. It cannot be explained by existing fundamentals.

Consciousness is universal. Every system has some degree of consciousness. This is called "panpsychism"—*pan* for all, *psych* for mind. Everything is conscious, not just animals but even particles. Physicist David Bohm published a paper titled "A New Theory of the Relationship of Mind and Matter," promoting a panpsychist theory of consciousness.

Representationalism and Tye's PANIC Theory

According to the representational theory of consciousness—or, for short, representationalism—conscious experiences can be explained in terms of the experiences' representational properties. Thus when I look at the mountain, my conscious experience of the mountain is just a matter of my experience's representation of the mountain. The phenomenal character of my experience is identified with its representational content.

The acronym PANIC stands for poised, abstract, nonconceptual, and intentional content. Therefore, for Michael Tye, professor of philosophy at the University of Texas, a mental representation qualifies as conscious only when its representational content is intentional, nonconceptual, abstract, and poised. Tye holds that at least some of the representational content in question is nonconceptual (N), which is to say that the subject can lack the concept for the properties represented by the experience in question, such as an experience of a certain shade of red that one has never seen before. Conscious states clearly must also have "intentional content" (IC) for any representationalism. Tye also asserts that such content is "abstract" (A) and not necessarily about particular concrete objects. This condition is needed to handle cases of hallucinations where there are no concrete objects at all or cases where different objects look phenomenally alike. Perhaps most important for mental states to be conscious, however, is that such content must be "poised" (P), which is an importantly functional notion. The key idea is that experiences and feelings make a direct impact on beliefs and/or desires. For example, feeling hungry has an immediate cognitive effect, namely the desire to eat. States with nonconceptual content that are not so poised lack phenomenal character because they arise too early, as it were, in the information processing.

My argument against representationalism is that two possible experiences with different phenomenal properties have the same representational properties. Suppose you stand in the middle of an empty road. All you can see are two trees. The two trees, A and B, have the same size and shape, but A is twice as far from you as B is. Being aware that the two trees have the same size, you represent to yourself that they have the same properties, yet B takes up more of your visual field than A in a way that makes you experience the two trees differently. There is phenomenal difference without representational difference.

Higher-Order Monitoring Theory

According to higher-order monitoring theory (HOMT), what makes the mental state conscious is the fact that the subject is aware of it in the right way. According to HOMT, a mental state M of a subject S is conscious when, and only when, S has another mental state M*, such that M* is an appropriate representation of M. What provides conscious status of M is something outside M, namely M*. Neither M nor M* can act independently of the other state. It is their coming together in the right way that yields consciousness.

There is distinction between attentive awareness and inattentive awareness. Right now I am visually (focally) aware of my computer, but I am also visually (peripherally) aware of the ashtray at the far corner of my desk. The former is an attentive form of awareness, and the latter is an inattentive form of awareness. A similar distinction applies to all kinds of awareness.

Rosenthal's Higher-Order Thought Theory

According to David Rosenthal, a mental state has consciousness when its subject has a suitable higher-order thought about it. A conscious mental state M, of mine, is a state that is actually causing a nonconscious belief that I have M and causing it a noncomparable manner. An account of phenomenal consciousness (i.e., experienced as they are perceived and how they appear) can then be generated by stipulating that the mental state M should have a causal role and/or content of a certain distinctive sort in order to count as an experience. Additionally, when M is an experience of a mental image, bodily sensation, or emotional feeling, it will be phenomenally conscious when, and only when, suitably targeted.

A phenomenally conscious mental state is a state of a certain sort with nonconceptual intentional content, which is the object of a higher-order thought, causing that thought noninferentially. The two most familiar forms of higher-order theory postulate the existence of a pair of distinct mental states: a first-order perceptual or quasi-perceptual state with a given content and a higher-order thought or perception representing the presence of that first-order state, thereby rendering it conscious.

Cognitive Theory of Consciousness

Cognition is the mental action or process of acquiring knowledge and understanding through thought, experience, and the senses. Cognitive theory is an approach to psychology that attempts to explain human behavior by understanding the thought processes. Here, consciousness is treated as an experimental variable to look for general capacities that distinguish conscious and unconscious mental functioning. Newton came up with his famous theory of gravity by treating gravity as a variable by considering presence and absence of gravity. In the same way, breakthrough in consciousness will occur if we treat it as a variable. Conscious processes are phenomenally serial, internally consistent, unitary at any moment, and limited in capacity. Nonconscious processes are functionally concurrent, highly differentiated from each other, and relatively unlimited in capacity when taken together.

Bernard Baars identified eight psychological functions of consciousness as follows:

definition and context setting

adaptation and learning

prioritizing and access control

recruitment and control of thought and action

decision-making and executive function

error detection and editing

reflection and self-monitoring

optimizing and trade-off between organization and flexibility

Each proposed function of consciousness is served by interplay of conscious and unconscious processes.

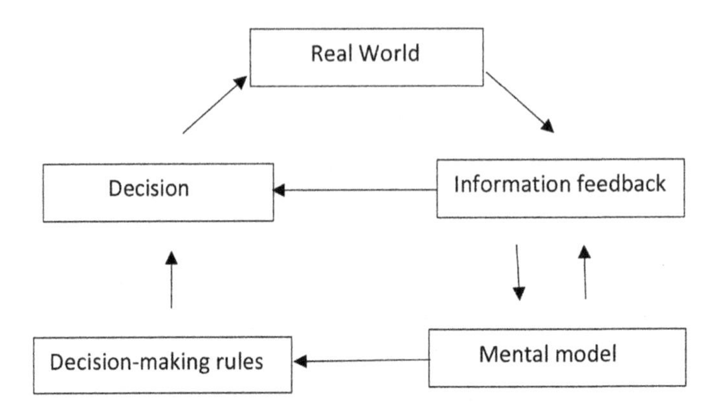

Figure 6: Mental model

A mental model (figure 6) is a kind of internal symbol or representation of external reality, hypothesized to play a major role in cognition, reasoning, and decision-making. It is an explanation of someone's thought process about how something works in the real world. It is a representation of the surrounding world, the relationships between its various parts and a person's intuitive perception about his or her own acts and their consequences. Figure 6 shows how information gathered from the real world goes to the mental model that makes decisions according to decision-making rules. Then it goes back to the real world.

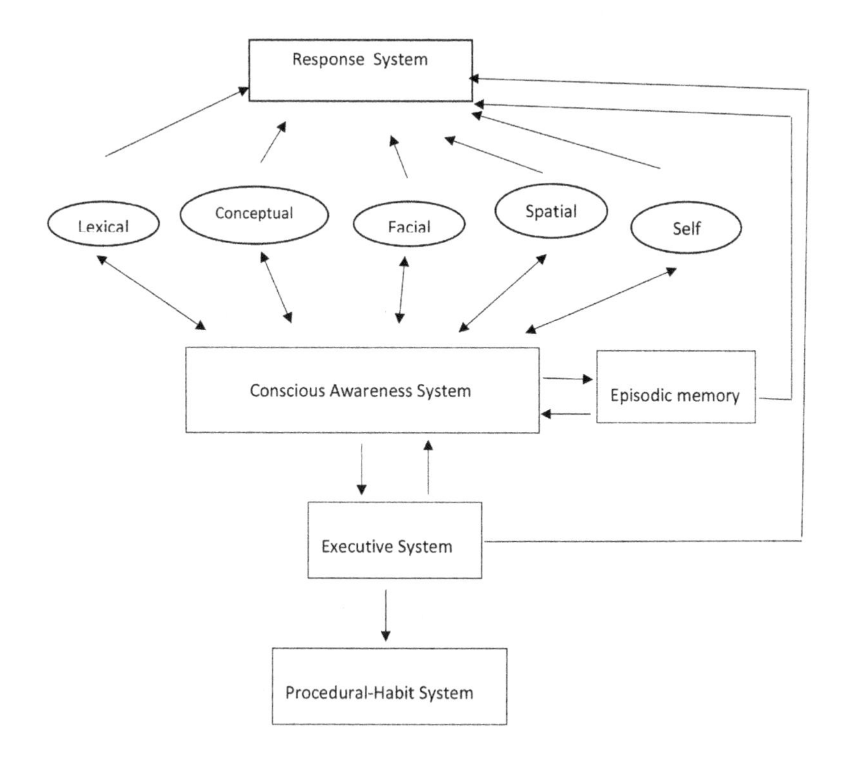

Figure 7: Schacter's DICE model

Daniel L. Schacter proposed his dissociable interactions and consciousness experience (DICE) model (figure 7). In Schacter's DICE model, consciousness operates in the consciousness awareness system. DICE gives consciousness an information-processing role in integrating the output of specialized modules and transmitting them to an executive system by interconnections between the conscious awareness system and individual awareness system and episodic memory. The DICE model suggests that the primary role of consciousness is to mediate voluntary action under the control of an executive.

Views of Other Researchers

Excerpted below are the views of other prominent researchers on consciousness.

Antonio Damasio

His three-layered theory of consciousness is based on a hierarchy of stages, with each stage building upon the last. The most basic representation of the organism is referred to as the proto-self. Next is core consciousness, and finally, extended consciousness.

Ned Joel Block

He makes a distinction between phenomenal consciousness and access consciousness. Phenomenal consciousness consists of subjective experience and feelings. Access consciousness is available in the cognitive system for the purposes of reasoning, speech, and high-level action control.

Jeffrey W. Cooney and Michael S. Gazzaniga

They propose that a neural workplace created by the presence of a large-scale network interconnects the outputs of numerous specialized brain regions and integrates providing a promising solution.

S. Dehaene and L. Naccache

They propose global neural workplace. According to them, attention is a prerequisite for consciousness. However, considerable amount of neural processing is possible without consciousness. Intentional behavior, cognitive tasks, and novel operations require consciousness. They propose a global neural workspace. Many cerebral networks are active in parallel and process information unconsciously. If large neurons in the brain are mobilized by top-down attention into coherent activity, then the information becomes conscious.

Daniel Dennett

Dennett put forward a theory that there is no single central place where conscious experience occurs; instead, there are various events of content-fixation occurring in various places at various times in the brain. The brain consists of a bundle of semi-independent agencies. When content-fixation takes place in one of these, its effects may propagate so that it leads to the utterance of one of the sentences that make up the story in which the central character is one's "self." Dennett's view of consciousness is that it is the apparently serial account for the brain's underlying parallelism.

Gerald Maurice Edelman and Giulio Tononi

Their dynamic core hypothesis proposes that the neural mechanisms underlying consciousness consist of a functional cluster in the thalamocortical system (i.e., nuclei in the thalamus and neurons in all modular areas of the cortex), within which reentrant neuronal interactions yield a succession of differentiated yet unitary metastable states. The boundaries of the dynamic core are suggested to shift over time, with some neuronal groups leaving and others being incorporated, these transitions occurring under the influence of internal and external signals. A notable feature of the dynamic core hypothesis is the proposal of a quantitative measure of "neural complexity," high values of which are suggested to accompany consciousness.

Anthony Freeman

According to Freeman, activity patterns created by the sensory dynamics are spread out over large areas of cortex, not concentrated at points. The outflow also is widely distributed. Consciousness is an emergent property of the brain, resulting from the communication of information across all its regions, and cannot be reduced to something residing in specific areas that control for qualities like attention, hearing, or memory. Different regions modulate different characteristics of consciousness, such as attention, language, and self-control, but none of these qualities are sufficient on their own to create what we experience as consciousness.

Nancy Kanwisher

The contents of awareness are not represented in a single unitary consciousness system, but rather that each conscious perceptual content is represented in the same set of neurons that analyze that perceptual information in the first place. Further, there is now compelling evidence from several different techniques showing that perception without awareness is possible. Thus a strong neural representation in a given cortical area is not sufficient for awareness of the information so represented, raising the question of which perceptual information will reach awareness. I speculate that in order for a focal neural representation to reach awareness it may have to be accessible to other parts of the brain. The construction of a fully conscious percept may involve interactions between domain-specific systems for representing the contents of awareness (primarily in the ventral visual pathway) and domain-general systems (primarily in the dorsal pathway) for organizing those contents into structured percepts.

Rodolfo Llinas

Llinas uses the term *qualia* to denote subjective experience of any type generated by the nervous system, be it pain, the color green, or the specific timber of a musical note. Essentially, the question of qualia or feelings is the question of conscious experience. The approach when considering consciousness is to imagine its hierarchically as a special province of the human brain. Other creatures have varying, albeit lesser, forms of experience, but nothing that compares to the rich language-infused complexity of the human mind. Nevertheless, such judgments are merely a reflection of our capacity to analyze, compare, and elevate the primacy of our own experience. Llinas suggests that consciousness, or subjective experience, is not an exclusively human capacity but stems from neural activity itself.

Geraint Rees

The directness and vivid quality of conscious experience belies the complexity of the underlying neural mechanisms, which remain incompletely understood. Recent work has focused on identifying the brain structures and patterns of neural activity within the primate visual system that are correlated with the content of visual consciousness. Functional neuroimaging in humans and electrophysiology in awake monkeys indicate that there are important differences between striate and extrastriate visual cortex in how well neural activity correlates with consciousness. Moreover, recent neuroimaging studies indicate that in addition to these ventral areas of visual cortex, dorsal prefrontal and parietal areas might contribute to conscious visual experience.

John Searle

John Searle defines consciousness by its four features—it's real and irreducible, caused by brain processes, exists in the brain, and functions causably. He argues for a biological understanding that counters many of the philosophical conceptions. Information about consciousness is distributed to many neuronal population distributed throughout the brain.

Francisco Varela

Neurophenomenology refers to a scientific research program aimed to address the hard problem of consciousness in a pragmatic way. It combines neuroscience with phenomenology in order to study experience, mind, and consciousness, with an emphasis on the embodied condition of the human mind. Varela, in particular, puts these analyses to use in his neurophenomenological approach to consciousness and offers a neurophenomenological account of time consciousness as "an acid test of the entire neurophenomenological enterprise."

Globalist view of consciousness

The globalist view of consciousness is summarized below.

> The architecture of consciousness comprises numerous semiautonomous specialist systems, which interact in a dynamic way in a global workspace.
>
> The global workspace recruits resources by global distribution of information.
>
> Specialist systems compete for access to global workspace.
>
> Access to the workspace is gated by a set of active context and goals.

According to the globalist view, consciousness provides a momentary unifying influence for a complex system through global access and global distribution.

Concluding Remarks

Neuroscientists have been trying to explain that consciousness is created by the neural activities of the brain, but they have not proved yet which part of the brain creates consciousness and how it does it. Neuroscience needs a theory of consciousness that explains what the phenomenon is, as well as what kinds of entities possess it. Two important theories have attempted to address this: integrated information theory (IIT) by Giulio Tononi, helped by Christof Koch, and global workspace theory (GWT) by Bernard Baars. Other researchers have been exploring cognitive theory to explain consciousness.

According to David Chalmers, consciousness is fundamental. Physicists sometimes take some aspects of the universe as fundamental building blocks: space, time, and mass, with fundamental laws like the laws of gravity or of quantum mechanics. His second idea is panpsychism, proposing that consciousness might be universal, with every system having some degree of consciousness.

Tye proposed representationalism with PANIC theory, stating that mental representation qualifies as conscious only when its representational content is intentional, nonconceptual, abstract, and poised. Higher-order monitoring theory distinguishes between attentive form of awareness and inattentive form of awareness. Higher-order thought theory postulates the existence of a first-order perceptual or quasi-perceptual state with a given content and a higher-order thought or perception representing the presence of that first-order state, thereby rendering it conscious.

Cognition is a mental action and process of acquiring knowledge or understanding through thought, experience, and senses. Bernard Baars proposed function of consciousness as interplay of conscious and unconscious processes. Schacter proposed the DICE model, where consciousness operates in the consciousness awareness system as an information-processing role. In addition, we have discussed the views of other prominent researchers on consciousness. However, not all these theories explain consciousness itself—but rather a causal basis of consciousness. Although neuroscientists are working hard to prove that the brain generates consciousness, there is no such proof yet. Therefore, there is no decision between monism and dualism of consciousness. Another research area where progress can be made in understanding consciousness is by tightening the correlation between the theoretical and experimental basis of consciousness.

In the next chapter, we will discuss consciousness based on quantum mechanics.

CHAPTER 5

Quantum Consciousness

There is a growing agreement among scientists that consciousness is unlikely to arise from classical properties of matter. The more we know about the structure and function of brain, the less we are able to explain how consciousness can arise from classical properties. The idea that consciousness is a manifestation of a complex net of electric pulses within the brain is now discredited. However, quantum theory provides a new theory of matter altogether. Quantum physics has been particularly intriguing for scientists to provide a physical explanation of consciousness.

Newtonian classical physics failed to explain the behavior of matter at the micron level, like electron, proton, and so on. This is where quantum physics came into play. Since neuron dimensions are of a micron level, their behavior cannot be explained by classical physics. Since the classical physics they are using do not work in the microworld, that is why many neurobiologists reach the conclusion that classical physics cannot explain consciousness. In this chapter, we will describe the research work that has been ongoing to explain consciousness by quantum physics. But first we will briefly explain Newtonian classical physics and quantum physics.

Classical Physics

Classical physics started from Isaac Newton, the greatest creative genius science had ever seen before Einstein. Newton's great masterpiece *Principia* has probably had a greater influence in our civilization than any other book except the Bible. There is a popular story that Newton was sitting under an apple tree; an apple fell on his head, and he suddenly thought of the universal law of gravitation. As in all such legends, this is almost certainly not true in its details. Probably the more correct version of the story is that Newton, upon observing an apple fall from a tree, began to think along the following lines: the apple is accelerated as its velocity changes from zero as it is hanging on the tree and moves toward the ground. Thus there must be a force that acts on the apple to cause this acceleration. He called this force gravity.

In his later years, Newton was asked how he arrived at his theory of universal gravitation. "By thinking on it continually," was his response. "Continual thinking," for Newton, was almost beyond mortal capacity, with nonstop passion and obsession, living without food or sleep, on the edge of a breakdown. Newton's three laws of motion provide relationships between the motion of a body and the forces acting on the body. The first law states that if a body is at rest, it will be resting unless acted upon by an external force. Thus, a stone in the garden will be there forever unless an external force, caused by an event, moves it. The second law states that if an external force acts upon a body, the body will accelerate in the direction of that force. The acceleration will be proportional to the force and inversely proportional to the mass of that force. If a truck and a car each stalled on a level road, it would take many more people to push the truck—that is, to accelerate it from rest—than to get the car moving at the same velocity. This is because the truck has a greater mass than the car and the force required is directly proportional to the mass to have certain acceleration. The third law states that the reaction is equal in magnitude and opposite in direction. Thus when I sit on a chair, I am pressing my weight on the chair. The chair pushes back with equal force, and I sit comfortably. If I push harder than the chair can push back, the chair will collapse. If the chair were to push back harder than I push on it, then I would fly into the air. By the third law, equilibrium is reached and no harm is done.

Newton postulated that light is composed of particles, or corpuscles, and investigated the refraction of light, demonstrating that a prism decomposes white light into a spectrum of colors and that a lens and a second prism could recompose the multicolored spectrum into white light. He also showed that the colored light does not change its properties, regardless of whether it is reflected, scattered, or transmitted. He invented a reflecting telescope (known as a Newtonian telescope today). Newton postulated the existence of ether as a media to transmit light.

Wave Theory of Light

Newton had difficulty explaining diffraction and interference of light with his corpuscular theory of light. Diffraction is the slight bending of light as it passes around the edge of an object. Interference is the pattern developed due to superposition of two waves. This was solved by English scientist Thomas Young, who postulated a wave theory of light by his famous double-slit experiment, demonstrating that light does travel in waves.

In the early nineteenth century, scientists discovered connections between electricity and magnetism. An electric current flowing through a wire produces a magnetic field around it. The reverse is also true. Physicist James Clerk Maxwell showed mathematically that electric and magnetic fields travel through space in the form of waves and at the constant speed of light. The traveling electric and magnetic waves are called electromagnetic waves.

Special Theory of Relativity

Albert Einstein, considered the greatest physicist of all time, laid the foundations of much of twentieth-century physics. Einstein's first postulate states that all observers moving at constant speeds, even if those are different from each other, must witness the identical laws of physics. All speeds are relative; absolute speeds do not exist. Thus when we are traveling in a train, we find that the train is moving relative to the houses or trees by the side of the rail track. If we just look at another person or picture inside the train, we would not see that the train is moving. This is the first principle of relativity. Einstein's second postulate states that the speed of light is always measured as the same by all observers, independent of their own motion or the motion of the emitting body. Thus, the speed of light in a vacuum is a universal constant c of value 2.9979×10^8 meters per second. Light is an electromagnetic wave. I have summarized the postulates below.

Observers in constant speeds (inertial frames), having no acceleration, witness identical laws of physics. The speed of light is constant, independent of the motion of the observer or the body emitting the light. This is called Einstein's theory of special relativity. This has created the following effects:

> Time dilation: This is not noticeable in normal life. However, as the speed of an object approaches the speed of light, the effect becomes enormous. Consider the following example: Sam and Bob are twins. Sam is an astronaut and goes on a space trip with a time clock. Bob and his time clock stay on Earth. When Sam returns to Earth, he is younger than Bob and his time clock reads an earlier time than Bob's clock. This means that Sam has aged less than his twin brother, Bob. Now, if Sam travels in the space module at nearly the speed of light and returns from a distant planet after few months in his time measurement, he might find that his great-grandchildren have died of old age. A real paradox!
>
> Contraction of length: For everyday situation, the velocity is so much smaller than the speed of light that the length contraction is not noticeable and can be totally ignored. However, as the velocity approaches the speed of light, the length contraction is quite noticeable.

Increase of mass: When the object is moving, the mass increases, but slightly, at ordinary speed. However, as its speed approaches the speed of light, its mass increases significantly.

General Theory of Relativity

Einstein proposed another revolutionary theory: the mass of a body is equivalent to a concentrated form of energy, according to his famous equation: $E = mc^2$; where E is the energy, m is the mass of a body, and c is the velocity of light. Since c^2 has an immense magnitude, this mass-equivalent energy is also immense. To get an idea of this immense energy, to supply the energy need of a city with a population of three million, it would need mass-equivalent energy of a body of mass thirty-three grams. The energy of an atomic bomb explosion comes from converting some mass directly into energy.

Special theory of relativity deals with bodies moving at a constant velocity and does not deal with acceleration. Einstein proposed another revolutionary theory call the general theory of relativity. In this theory, gravity is not really a force; it is a manifestation of the curvature of space/time (note: not space but space/time—this distinction is crucial). Now, what is space/time? If we toss a ball, it travels a short distance in space but an enormous distance in time, as one second equals about three hundred thousand kilometers in units where velocity of light $c = 1$. Thus, a slight amount of space/time curvature has a noticeable effect.

If we place a heavy object, such as a bowling ball, on a trampoline, it will produce a dent in the trampoline. This is analogous to a large mass such as Earth causing the local space/time geometry to curve. The larger the mass, the bigger the amount of curvature. A relatively light object placed near the dent, such as a Ping-Pong ball, will accelerate toward the bowling ball in a manner governed by the dent. Firing the Ping-Pong ball at a suitable combination of direction and speed toward the dent will result in the Ping-Pong ball's orbiting the bowling ball. This is analogous to the moon orbiting Earth.

Blackbody Radiation

A problem came with the classical physics in explaining blackbody radiation. In physics, a blackbody is an object that absorbs all electromagnetic radiation that falls onto it. No radiation passes through it, and none is reflected. This lack of both transmission and reflection is to what the name refers. A blackbody represents a system in which the thermal energy is carried via electromagnetic radiation. Hence, it is possible to approximate the temperature of the object through the wavelength of the electromagnetic radiation that is emitted. Blackbodies below around 430 degrees Celsius produce very little radiation at visible wavelengths and appear black (hence the name). Blackbodies above this temperature, however, produce radiation at visible wavelengths, starting at red and going through orange, yellow, and white before ending up at blue as the temperature increases.

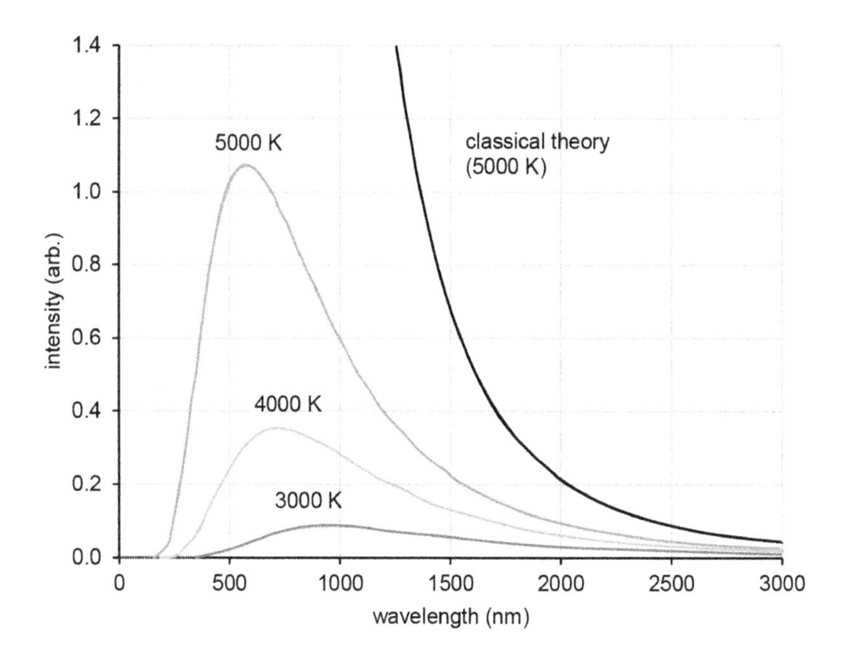

Figure 8: Blackbody curve at different temperatures

In the laboratory, a blackbody radiation is approximated by the radiation from a small hole entrance to a large cavity. Any light entering the hole would have to reflect off the walls of the cavity multiple times before it escaped, in which process it is nearly certain to be absorbed. This occurs regardless of the wavelength of the radiation entering (as long as it is small compared to the hole). The hole, then, is a close approximation of a theoretical blackbody, and if the cavity is heated, the spectrum of the hole's radiation (i.e., the amount of light emitted from the hole at each wavelength) will be continuous and will not depend on the material in the cavity.

The blackbody radiation at different temperatures is shown in figure 8. As can be seen, each curve has a definite peak. It passes through a maximum intensity at a certain frequency and then declines, but according to Rayleigh-Jeans law in classical physics, the intensity keeps on increasing with frequency, as shown in figure 8, for radiation at five thousand kelvins. This has been called ultraviolet catastrophe.

German physicist Max Planck solved the problem by introducing an equation for the radiation with a constant called Planck's constant, h. In classical physics, it was always assumed that the difference in energy radiated by two atoms could be as small as we chose. Planck's law provides a lower limit to that difference. Thus continuous gradation of energy is replaced by the discrete stepwise change of energy. Therefore, the energy emitted by the light particle photon is $E = hv$, where E is energy, h is Planck's constant, and v is frequency. The Planck constant has dimensions of energy multiplied by time and is expressed in joule seconds (J.s). The value of the Planck's constant is $h = 6.62606896 \times 10^{-34}$ J.s.

Planck calculated that the number of photons is fewer, as the energy level is higher. Hence, there are fewer photons at the ultraviolet frequency, solving the ultraviolet catastrophe.

Planck's constant helped solve the puzzle of photoelectric effect. When weak red light shines on a cesium surface, a few slow electrons are emitted. If the red light is made brighter, more electrons are emitted, but they still travel slowly. If a dim blue light is used, there are few electrons emitted, but they travel faster. If the blue light is made brighter, more electrons are emitted, but their velocities remain the same. If infrared light is used, no electrons are emitted.

Einstein solved this puzzle by assuming that the incident light was quantized, and light was made of particles called photons. He used the equation $E = h\nu$, where E is the energy of photon, h is the Planck's constant, and ν is the frequency of the incident light. Cesium atoms that are hit by photons emit electrons. Blue photons have more energy (hence, more velocity) than the red photons, as the blue light has higher frequency than the red light. From the viewpoint of classical physics, emitted electron velocity ought to increase with increasing incident light intensity, regardless of the light's frequency.

Quantum Mechanics

When an electron is orbiting around the nucleus, then the electrical force of attraction between the negatively charged electron and the positively charged proton in the nucleus will force the electron to collapse into the nucleus. Niels Bohr solved this problem by introducing discrete energy levels of the electrons by using Planck's constant. If the electron can change its state of motion in discrete steps, it must then stay in a particular orbit until it emits or absorbs enough energy in one single process to go from one orbit to another. This leads to discrete orbits, and transitions from one orbit to another give results in either the absorption or emission of discrete energy levels. Bohr's atomic theory successfully explained the atomic structure of hydrogen, but when it came to atoms of other elements, his theory was not sufficient. This is when quantum mechanics came into the picture.

The term *quantum* (Latin for quantity) refers to the discrete quantities that the theory assigns to physical items, such as the energy of light and electromagnetic wave. Quantum mechanics includes the following attributes:

1. Laws of probability
2. Discreteness of energy
3. Wave-particle duality
4. Schrödinger equation
5. Uncertainty principle
6. Exclusion principle
7. Spin of a particle
8. Quantum numbers

Laws of Probability

Probability is the likelihood that a certain thing will happen. The probability theory is used extensively in statistics, mathematics, science, and philosophy to draw conclusions about the likelihood of potential events and the mechanics of complex systems. The fundamental principle of quantum mechanics is based on the law of probability, not the law of certainty. Let us look at the picture of an atom in its ground state. If we think that the electron is looping around the nucleus, as shown in figure 9, then we are eighty years out of date. According to quantum mechanics, the probability density plot of the electron in an atom is shown in figure 10. The density of the dots represents the probability of finding the electron in that region.

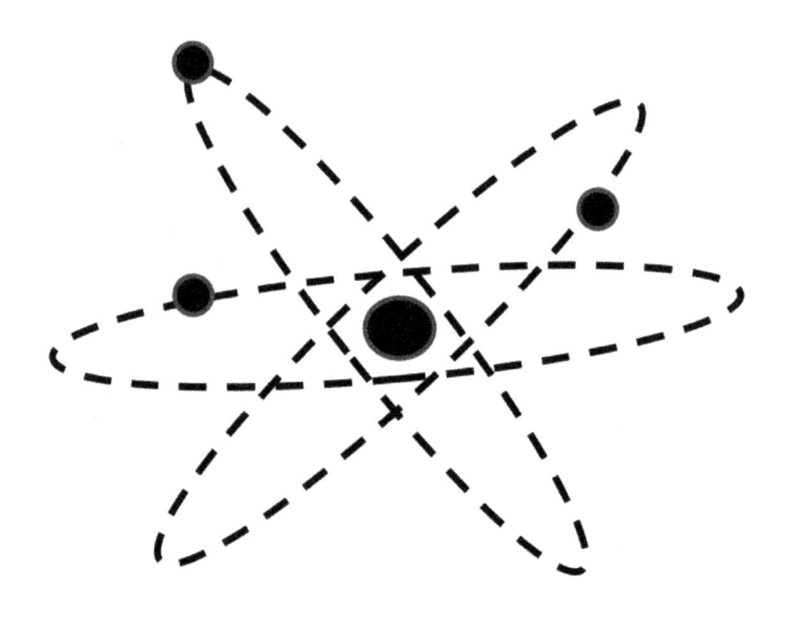

Figure 9: Electron looping around the nucleus according to classical physics

The size of an atom is ~1 x 10^{-10} meter. The central nucleus, where 99.9 percent of the atom's mass resides, is ~1 x 10^{-15} to 1 x 10^{-14} meter. The size of an electron is not exactly known, but the classical electron radius is thought to be of the order of 2.8 x 10^{-15} meter. If the nucleus were the size of our sun, then the hydrogen ground state would be twenty times larger than the solar system. If the electron were really a point particle moving around the atomic space, it would reside in a space so vacant that it would make the solar system seem crowded. Now, if we consider the wave function of the electron with its probability density, then the probability density will fill up the whole atomic space. For instance, try to push your hand through a wall. Because atoms are mostly empty space, their electrons are too small to stop you. But the probability density clouds of the atoms push your hand back. Effective, aren't they?

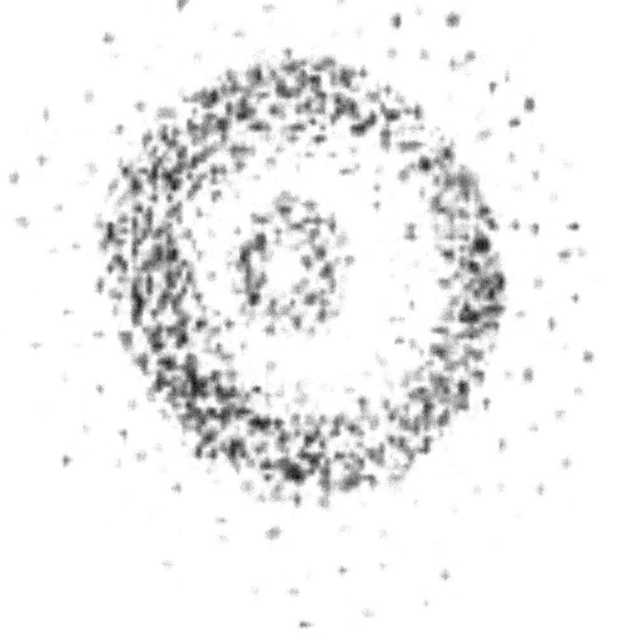

Figure 10: Electron probability density cloud

Discreteness of Energy

As discussed before, Niels Bohr introduced discrete energy levels of the electrons in atoms.

Wave-Particle Duality

Louis de Broglie hypothesized that all entities have both wave and particle aspects. De Broglie pointed out that a photon's wavelength, a wave property, could be related to the photon's momentum, a particle property, by combining $E = h\nu$ with another energy equation: $E = mc^2$.

As we discussed earlier, E is the energy, h is the Planck's constant, ν is the frequency, m is the mass, and c is the speed of light. Combining the two equations, we get $E = h\nu = mc^2$, from which we obtain $mc = h\nu/c$.

Because the photon's speed is c, the term mc in the equation above can be regarded as the photon's momentum p, hence $p = h\nu/c$.

Now, we know that $\lambda\nu = c$, which connects the wavelength λ, frequency ν, and speed of light c. Hence, $\nu = c/\lambda$. Substituting the value of ν, $p = h\nu/c = h/\lambda$.

This equation proves that the momentum (which relates to a particle) is inversely proportional to the wavelength (which relates to a wave), thus showing the wave-particle duality. This equation refers to photons, but de Broglie saw no reason that electrons and other particles of matter should not also have associated frequencies and wavelengths.

Clinton Davisson and Lester Germer confirmed the wave nature of electron, as hypothesized by de Broglie, from their experiment. Physicist Lawrence Bragg treated x-rays as waves and derived a famous equation in the analysis of an x-ray diffraction pattern. Davisson and Germer applied the same equation in explaining the diffraction pattern of electrons and confirmed the wave nature of electrons.

Schrödinger's Equation

Erwin Schrödinger invented a differential equation to derive wave functions for electrons. When the equation is solved, it yields a set of wave function represented by the Greek letter Ψ (psi). According to the equation, electrons confined in their orbits would set up standing waves and would describe the probability where an electron could be. For a single electron in a free atom, the wave function square Ψ^2 measures the probability of finding the electron at or near a given location. The larger the value of Ψ^2, the greater the probability of finding the electron at that location. The distribution of these probabilities forming regions of space around the nucleus are called orbitals. Thus orbitals could be called an electron density cloud. The densest area of the cloud has the greatest probability of finding an electron, and the least dense area has the lowest probability of finding an electron, as shown in figure 10.

Uncertainty Principle

Werner Heisenberg proposed the uncertainty principle for quantum mechanics, stating that the position and momentum of a particle cannot be accurately determined simultaneously. If the position of the particle is accurately known, then the speed and direction of the speed will have poor accuracy. This is because the measurement itself changes the position, speed, or direction of the particle. This is not obvious in the macroscopic world around us. If we want to measure the length of a table with a tape, the measurement does not change the length or position of the table in our acceptable measuring capacity. In the case of an electron, measurement becomes different. The act of looking at the electron with a super-high-power magnifier uses light made of photons. These photons would have enough momentum that they once hit the electron, it would change its course. Heisenberg wrote the uncertainty principle as $\Delta p \Delta x = \hbar$.

On the right side is the Planck's constant h divided by 2π (called reduced Planck's constant). Momentum is represented by p, and position (distance) is represented by x. The Δ symbols mean uncertainty. Hence, Δx is the uncertainty of position, and Δp is the uncertainty of momentum. Since h on the human scale is extremely small, Δx and Δp are, for all practical purposes, zero in the everyday world.

Exclusion Principle

The Pauli exclusion principle is a quantum mechanical principle formulated by Wolfgang Pauli. This principle is significant because it explains why matter occupies space exclusively for itself and does not allow other material objects to pass through it, while at the same time allowing light and radiation to pass. It states that no two identical fermions—electrons, protons, and neutrons, are fermions—may occupy the same quantum state simultaneously. For electrons in a single atom, it states that no two electrons can have the same four quantum numbers. The Pauli exclusion principle is one of the most important principles in physics, primarily because the three types of particles from which ordinary matter is made—electrons, protons, and neutrons—are all subject to it; consequently, all material particles exhibit space-occupying behavior.

The Pauli exclusion principle helps explain a wide variety of physical phenomena. One such phenomenon is the rigidity or stiffness of ordinary matter (fermions). The principle states that identical fermions cannot be squeezed into each other—material objects collide rather than passing straight through each other, and we are able to stand on the ground without sinking through it. Another consequence of the principle is the elaborate electron shell structure of atoms and the way atoms share electron(s) and a variety of chemical elements and their combinations (chemistry). An electrically neutral atom contains bound electrons equal in number to the protons in the nucleus. Because electrons are fermions, the Pauli exclusion principle forbids them from occupying the same quantum state, so electrons have to pile on top of each other within an atom.

Spin of a Particle

Atomic particles possess an intrinsic angular momentum. An electron has a magnetic field due to its spin. When electrons that have opposite spins are put together, there is no net magnetic field because the positive and negative spins cancel each other out. Experiments suggest just two possible states for this angular momentum, and following the pattern of quantized angular momentum, this requires an angular momentum quantum number of 1/2 with two states +1/2 and -1/2.

Quantum Numbers

There are four quantum numbers to specify the movement of electrons within an atom. Bohr's model used one quantum number to describe the distribution of electrons in the orbits of the atom. The only information required was the sizes of the orbits, which was described by the quantum number n. Schrödinger's model allowed the electrons three-dimensional space. Therefore, three coordinates or three quantum numbers were required. The three quantum numbers are the principal quantum number (n), angular quantum number (l), and magnetic quantum number (m). These quantum numbers describe the size (n), shape (l), and orientation (m) in space of the orbitals of electrons of an atom.

However, there are two electrons in an orbital. To distinguish between two electrons in an orbital, a fourth quantum number called spin quantum number (s) is required. One quantum number s is +1/2, and the other quantum number s is -1/2. It takes three quantum numbers (n, l, m) to define an orbital and fourth quantum s to identify an electron that can occupy the orbital.

Debate on Quantum Mechanics

Einstein remained unhappy with the standard probabilistic interpretation of quantum mechanics. Einstein to Bohr: "God does not play dice with the universe." Bohr to Einstein: "Stop telling God how to behave." While their actual exchange was not quite so dramatic and quick as the paraphrase would have it, this was nevertheless a wonderful rejoinder from what must have been a severely exasperated Bohr. The Bohr-Einstein debate had the benefit of forcing the creators of quantum mechanics to sharpen their reasoning and face the consequences of their theory in its most starkly nonintuitive situations. The foundations of the subject contain unresolved problems—in particular, problems concerning the nature of measurement. An essential feature of quantum mechanics is that it is generally impossible, even in principle, to measure a system without disturbing it. The detailed nature of this disturbance and the exact point at which it occurs are obscure and controversial. In spite of the indeterminacy, the physicists who explain the mysteries of atomic and subatomic particles are still pursuing quantum mechanics unhesitatingly.

Aspect, P. Grainger, and G. Roger, working at the University of Paris, created pairs of photons and sent members of each pair to detectors separated by a distance of thirteen meters. The detectors measured the polarization of the photons, a property related to their spin. This team showed that measuring the polarization of photons at one detector affected the polarization measured at the other detector. The influence that traveled between the detectors did so in less than ten nanoseconds. This was a quarter of the time less than a light signal would have taken to travel the thirteen-meter distance. If the team had the technology to measure an even smaller time interval, they could have found that this unknown influence traveled even faster. Hence, quantum mechanics is right and, alas, Einstein's paradigm is wrong. However, is there a signal that travels faster than light? In the experiment done by Aspect et al., the two photons were not separate in the first place. They had the same potential. In the world of potentiality, they were not two photons but just one potentiality with manifestation for both. In the potential world, they are united as one. Hence, they are interconnected beyond the limits of space and time. This is known as the theory of entanglement, which states that even though particles appear separated on the subatomic level, they in fact resonate together. Even if a particle is placed on one side of the universe and another is placed at the opposite side, they would resonate instantaneously when one is caused to vibrate.

Quantum Field Theory

The Schrödinger equation does not accommodate the special theory of relativity. Consider a particle moving at nearly the speed of light; we need special relativity, not quantum mechanics, to study its motion. Special relativity tells us that energy can be converted into matter. If the electron is energetic enough, an electron and a positron (antielectron) can be produced. The Schrödinger equation is simply incapable of describing such a phenomenon. This is where quantum field theory (QFT) comes in. QFT was born out of the necessity of accommodating special relativity and quantum mechanics, just as the new science of string theory is created for accommodating general relativity and quantum mechanics.

Quantum Approach to Consciousness

Since quantum theory is the most fundamental theory of matter currently available, it is a legitimate question to ask whether quantum theory can help us to understand consciousness. Quantum theory introduced an element of randomness standing out against the previous deterministic worldview. Quantum randomness in processes such as spontaneous emission of light, radioactive decay, or other examples of state reduction was considered a fundamental feature of nature, independent of our ignorance or knowledge. Other features of quantum theory, which were found attractive in discussing issues of consciousness, were the concepts of complementarity and entanglement.

Previously, the worldview of physics was a model of a great machine composed of separable interacting material particles. Schrödinger, Heisenberg, and their followers created a universe based on superimposed inseparable waves of probability amplitudes. This new view would be entirely consistent with the Vedantic concept of all in one. The unity and continuity of Vedanta are reflected in the unity and continuity of wave mechanics. Vedanta teaches that consciousness is singular, all happenings are played out in one universal consciousness, and there is no multiplicity of selves. Schrödinger believed that consciousness is never experienced in the plural, only in the singular. He viewed consciousness as fundamental to reality and nondual. Not only is there just one single consciousness, but that consciousness is not ultimately separate from objects experienced within consciousness.

Schrödinger was sympathetic to the Hindu concept of Brahman, by which each individual's consciousness is only a manifestation of a unitary consciousness pervading the universe, which corresponds to the Hindu concept of God. Schrödinger concludes, "I am the person, if any, who controls the 'motion of the atoms' according to the Laws of Nature."

Eugene Wigner proposed that potential must become actual when it appears in consciousness. Nothing physical—not the eye, not the brain—can project the state vector. Only a nonphysical consciousness can do it. The buck stops with consciousness. In order to account for the actual existence of anything physical, we are forced by Wigner's argument to recognize the existence of a nonphysical consciousness. The world cannot be just a bunch of inert matter, a collection of objects. There must also be a subject—a consciousness apart from objects—that is aware of them. However, Wigner was criticized for tying it to any particular person or object whatsoever. The subject is the one source of conscious awareness that shines forth through the many individuals. The object is the one potential for the appearance of the many phenomenal appearances in all the individual worlds. This reality is prior to space, time, and causality and is thus characterized by nonseparability and spontaneity.

The application of quantum mechanics to explain consciousness has three major technical problems. The first problem is that quantum mechanics is primarily a theory of atomic processes, whereas consciousness appears to be connected with macroscopic brain activities and processes.

The second problem is that quantum theory as developed for the study of atomic processes does not apply to biological systems such as the brain.

The third problem is that the Copenhagen interpretation of quantum theory regards the quantum formalism as merely a set of rules for calculating expectations about our observations, not as a description or picture of physical reality itself. However, without a description of physical reality, consciousness becomes a puzzle within an enigma.

Let us now explore how different scientists have explained quantum consciousness. The first detailed quantum model of consciousness was the synaptic tunneling model done by physicist Evan Walker. In synaptic model, electrons can tunnel between adjacent neurons, thereby creating virtual neural network overlapping the real one. This virtual nervous system creates consciousness and directs the behavior of the real nervous system. The real nervous system operates by means of synaptic messages. The virtual one operates by means of quantum tunneling—that is, particles passing through an energy barrier that classically they would not be able to climb. The real one is caused by classical laws, the virtual one by quantum laws, even if classical laws can describe the brain's behavior.

Physicist Nick Herbert proposed that consciousness is a pervasive process in nature. Mind is a fundamental component of nature as elementary particles and forces. Three features of quantum theory can detect mind:

Randomness.

Objects have attributes only when they are observed.

Once two particles have interacted, they remain connected.

According to Herbert, these three features can account for the features of mind: free will, essential ambiguity, and deep psychic connectedness.

Scientist James Culbertson has speculated that consciousness may be a relativistic feature of space-time. In his opinion, consciousness permeates all of nature so that every object has a degree of consciousness.

Quantum Brain Dynamics

Umezawa and Takahashi applied QFT to brain dynamics related to consciousness and called it quantum brain dynamic (QBD). Brain confined within the cranium is a coupled matter-radiation system composed of atomic constituents and their radiation fields, especially electromagnetic fields. According to QFT, matter made of atomic constituents can be described as the spatial distribution of quantum electric dipoles in a spatial region—that is, a quantum electric dipole field. Thus, brain is a quantum electric dipole field. Biomolecular structures are represented in the quantum electric dipole field by singularities, topological defects, local symmetries, and localizations of the field. Biomolecular architecture provides geometric objects emerging in the quantum electric dipole field. Brain tissue can be seen as the quantum electric dipole field equipped with the highly systemized geometric objects manifesting various local symmetries and breaking global symmetries—that is, breaking uniformity of the field.

Umezawa and Takahashi presented an interesting physical process for memory retrieval in terms of QFT. As long as memory is maintained in the form of geometric objects of the quantum electric dipole field, new quanta called Nambu-Goldstone bosons emerge from geometric objects triggered by arbitrarily small incoming energy. Emergence of Nambu-Goldstone bosons is memory retrieval. These Nambu-Goldstone bosons are transformed into quanta of electromagnetic field: photons. However, these are specialized photons in the sense that they have nonzero mass, do not propagate, and remain nearby the geometric objects.

Such a nonpropagating photon of the electromagnetic field is called a tunneling photon. Therefore, we may call photons surrounding the geometric objects of biomolecular architecture "biological tunneling photons." The mass of this biological tunneling photon is about ten electron volts, which is far smaller than the mass of an electron. The critical temperature for boson condensation of biological tunneling photons of mass about ten electron volts turns out to be actually higher than body temperature.

Memory retrieval of in terms of Nambu-Goldstone bosons can emerge from the biomolecular architecture of the geometric objects of the brain's quantum electric dipole field. Modern quantum field theory indicates that boson condensation of biological tunneling photons can occur at body temperature. Therefore, it seems plausible that the physical correlation of conscious mind might be this boson condensation of tunneling photons manifesting quantum coherence through the entire brain.

Neumann Approach

John von Neumann argued that the mathematics of quantum mechanics allows for the collapse of the wave function to be placed at any position in the causal chain, from the measurement device to the "subjective perception" of the human observer. In his "von Neumann chain," he starts with a quantum object, an observable of which is to be measured. However, based on the formalism of quantum theory and the Schrödinger dynamics in particular, because of the interaction between the object and the measuring instrument, the object is entangled with the instrument. Von Neumann extends this chain up to an observer, but if we take an observer into consideration, we simply end up with a description according to which the body of the observer, including his or her brain, is entangled with the instrument and the object. The measurement problem can then be further refined as to how it is that a state reduction to one of the eigenstates of the measured observable can occur in this chain. Von Neumann showed that as far as final results are concerned, you can cut the chain and insert a collapse anywhere you please. He felt that the process by which a physical signal in the brain becomes an experience in the human mind or human consciousness is the site of the wave function collapse.

The dynamics of the entire universe, including the bodies and brains of the conscious human participants/observers, is represented by three processes. Process 1 is the choice of the experimenter how he will act. This process is sometimes called the "Heisenberg choice" because Heisenberg emphasized its crucial role in quantum dynamics. Von Neumann does not specify the causal origins of this choice, apart from the conscious intentions of the human agent. Process 2 has the effect of expanding the microscopic uncertainties demanded by the Heisenberg uncertainty principle into the macroscopic domain. This conflict with conscious experience is resolved by invoking process 3, which is sometimes called the "Dirac choice." Dirac called it a choice on the part of nature. It can be regarded as nature's answer to a question effectively posed by process 1 choice made by the experimenter. Processes 1 and 3 act on the variables that specify the body/brain of the agent.

Penrose-Hameroff Approach

Penrose and Hameroff have worked extensively on quantum consciousness and have proposed a model called orchestrated objective reduction (Orch OR model). They suggested that quantum vibrational computations in microtubules were "orchestrated" ("Orch") by synaptic inputs and memory stored in microtubules and terminated by Penrose "objective reduction" ("OR"), hence "Orch OR."

Microtubules

Proteins called microtubules are best suited for quantum computation and objective reduction. Proteins are large biomolecules, or macromolecules, consisting of one or more long chains of amino acid residues. Ionic forces, hydrogen bonds, and dipoles are types of forces operating among amino acid side groups within a protein. Dipole-dipole interactions include three types, which are known as van der Waals forces and (1) permanent dipole to permanent dipole, (2) permanent dipole to induced dipole, and (3) induced dipole to induced dipole. Induced dipole to induced dipole interactions are the weakest and are known as London dispersion forces.

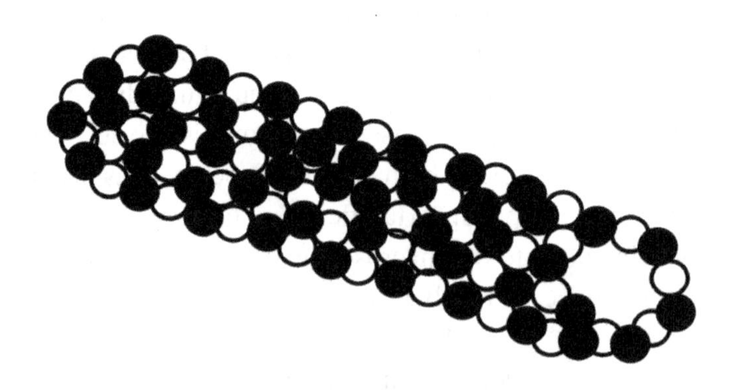

Figure 11: Structure of microtubule

Proteins might be using quantum superposition ("qubits") in determining their conformational states (bits). The possibility could be by using microtubule protein tubulin. The structure of microtubule is shown in figure 11. Tubulins are peanut-shaped dimers with two connected monomers, and they undergo several types of conformational changes. A dimer is a macromolecular complex formed by two usually non-covalently-bound macromolecules such as proteins or nucleic acids. A heterodimer is formed by two different macromolecules. Tubulin heterodimers assemble into linear protofilaments. Protofilaments in turn assemble into microtubules. All such assembly is subject to regulation by the cell. In neurons, microtubules self-assemble to extend axons and dendrites and form synaptic connections; microtubules then help maintain and regulate synaptic strengths responsible for learning and cognitive functions.

Dipole couplings among neighboring tubulins in the microtubule lattice act as "transition rules" for simulated microtubule automata exhibiting information processing, transmission, and learning. Classical microtubule automata switching in the nanosecond scale offer a potentially huge increase in the brain's computational capacity. Conventional approaches focus on synaptic switching that contain roughly 10^{11} brain neurons with 10^{3} synapses per neuron and switching in the millisecond range of 10^{3} operations per second. That predicts about 10^{17} bit states per second for a human brain. As the human brain contains about 10^{11} neurons, nanosecond microtubule automata offer about 10^{28} brain operations per second.

Objective Reduction (Orch OR)

Nevertheless, this vast computational complexity will not by itself address the difficult issues related to consciousness. Penrose proposes objective reduction (Orch OR) with quantum coherent states and quantum computation to explain consciousness. Penrose has proposed an objective threshold ("objective reduction"—OR) related to an intrinsic feature of fundamental space-time geometry. The objective factor in OR is an intrinsic feature of space-time itself (quantum gravity). Penrose begins from general relativity with the notion that mass is equivalent to space-time curvature. He concludes that quantum superposition—actual separation (displacement) of mass from itself—is equivalent to simultaneous space-time curvatures in opposite directions, causing "bubbles" or separations in fundamental reality. Penrose views the bubbles as unstable, with a critical objective degree of separation resulting in instantaneous reduction to classical unseparated states. Objective reductions are therefore events that reconfigure the fine scale of space-time geometry. Modern panpsychists attribute proto-conscious experience to a fundamental property of physical reality. If so, consciousness might involve self-organizing OR events rippling through an experiential medium.

Penrose's objective reduction is related to quantum gravity by the uncertainty principle by $E = h/(2\pi T)$, where h is Planck constant and E is the gravitational self-energy of the superposed mass (displaced from itself by, for example, the diameter of its atomic nuclei); and T is the coherence time until OR self-collapse. The size of an isolated superposed system is thus inversely related to the length of time until self-collapse. Large superposed systems (one kilogram) would self-collapse (OR) in only 10^{-37} seconds; an isolated superposed atom would undergo OR only after 106 years! If OR events occur in the brain coupled to known neurophysiology, then we can estimate that T for conscious OR events may be in a range from ten to five hundred milliseconds. This range covers neurophysiological activities such as twenty-five milliseconds "coherent forty hertz," one-hundred-millisecond EEG rhythms, and five-hundred-millisecond sensory perceptions. OR events coupled to roughly one-hundred-millisecond activities would require a few nanograms of superposed mass. Microtubules are best suited for quantum computation and objective reduction.

Microtubule subunit tubulins undergo coherent excitations. The coherent excitations are proposed to "clock" computational transitions occurring among neighboring tubulins acting as "cells," as in molecular-scale cellular automata. Dipole couplings among neighboring tubulins in the microtubule lattice act as transition OR rules for simulated microtubule automata exhibiting information processing, transmission, and learning. An Orch OR event is shown in figure 12. Classical computing in microtubilin simulation leads to emergence of quantum coherent superposition and quantum computing. A conscious event (Orch OR) occurs when self-collapse (Orch OR) happens when the mass energy meets critical threshold related to quantum gravity in coherence with other microtubule tubulins. The area under the curve in figure 12 shows mass energy with collapsed time T in accordance with $E = h/(2\pi T)$.

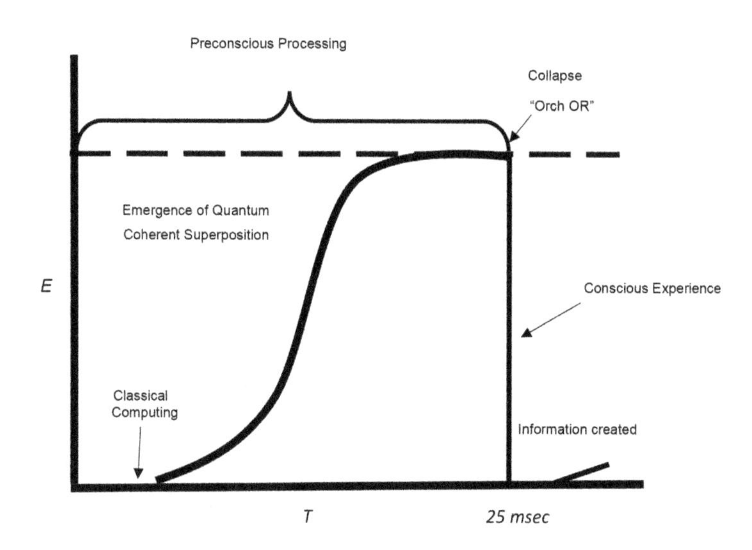

Figure 12: An Orch OR event

David Bohm Approach

Bohm evolved a new and controversial theory of the universe. It is a new model of reality that Bohm calls the "implicate order." Bohm's basic assumption is that elementary particles are actually systems of extremely complicated internal structure, acting essentially as amplifiers of information contained in a quantum wave. The theory of the implicate order contains an ultra-holistic cosmic view; it connects everything with everything else. In principle, any individual element could reveal detailed information about every other element in the universe. The central underlying theme of Bohm's theory is the unbroken wholeness of the totality of existence as an undivided flowing movement without borders.

Bohm parallels the activity of consciousness with that of the implicate order in general. Bohm conceives of consciousness as more than information and the brain; rather, it is information that enters into consciousness. Bohm describes it in terms of a series of moments. One moment gives rise to the next, in which context that was previously implicate is now explicate, while the previous explicate content has become implicate. Consciousness is an interchange; it is a feedback process that results in a growing accumulation of understanding.

Bohm considers the human individual to be an intrinsic feature of the universe, which would be incomplete in some fundamental sense if the person did not exist. Bohm believes that the individual who uses inner energy and intelligence can transform humankind. The collectivity of individuals has the potential to reach the principle of the consciousness of humankind. Continuing with this theme on the transformation of consciousness, Bohm goes on to suggest that an intense heightening of individuals who have shaken off the pollution of the ages (wrong worldviews that propagate ignorance), who come into close and trusting relationship with one another, can begin to generate the immense power needed to ignite the whole consciousness of the world. In the depths of the implicate order, there is a consciousness deep down of the whole of humankind.

It is this collective consciousness of humankind that is truly significant for Bohm. It is this collective consciousness that is truly one and indivisible. Each human person should contribute toward the building of this consciousness of humankind. There's nothing else to do; there is no other way out. That is absolutely what has to be done, for nothing else can work.

Kozlowski Approach

Miroslow Kozlowski and Janina Marciak-Kozlowska showed that the spectrum of the brain photons can be analyzed with the help of the Planck formula for equilibrium thermal radiation. They calculated the normalized energy spectra of the brain and the cosmic background photons. They showed that both spectra can be calculated with the same formula (Planck blackbody formula) with different temperatures. For cosmic background, the temperature is $T = 2.53 \times 10^{-4}$ eV = 2.93 K, and for brain photons, it is $T = 7.8 \times 10^{-15}$ eV = 9×10^{-11} K.

In another article, Kozlowski and Marciak-Kozlowska proposed a strong connection between the structure of the universe and brain activity. They showed that the predominant frequency of the brain vibration $f \simeq 12$ Hz can be achieved for oscillator with mass of the order of Planck mass. They showed that the Planck vibration and the brain vibration can be achieved by the same second-order Schrodinger equation.

Stapp Approach

Heisenberg believed that the probability distribution that occurs in quantum theory exists in nature herself. He considered that the evolution of this probability distribution was punctuated by uncontrolled quantum wave collapses, which are the events that actually occur in nature and the manifestation of which eliminates the other possibilities in the probability distribution. The things or events are controlled by quantum jumps that do not individually conform to any natural law but collectively conform to statistical rules.

With respect to the brain and consciousness, Stapp considers that brain processes involve chemical processes and hence should be treated quantum mechanically. He treats brain processes occurring at the synaptic junction triggered by the capture of calcium ions quantum mechanically. A quantum mechanical component must be added to the other uncertainties, such as those generated by thermal noise that enter into the decision as to whether or not the synapse will fire. Nonlinearity of the synaptic system and the large number of metastable states into which the brain can evolve point to a quantum mechanical structure. Although Stapp thinks there is quantum-based activity in the synapses and possibly other aspects of the brain, his theory, in contrast to quantum brain dynamics, or Orch OR, is not actually based at the microscopic level. Instead, Stapp envisages consciousness as exercising top-level control over neural excitation in the brain. Quantum brain events are suggested to occur at the whole brain level rather than the level of the synapses.

In this system, conscious events are selected from the large-scale excitation of the brain. Stapp views the brain as a self-programming computer with self-sustaining input from memory, which is a code derived from previous experience. This results in a number of probabilities from which consciousness has to select. The conscious act is a selection of a piece of top-level code, which then exercises control over the flow of neural excitation. Each human experience is accompanied by the activation of a top-level code.

According to Stapp, the top level of brain processing involves information gathering, planning of actions, choice of particular plans, and execution and monitoring of these plans. He suggests that each top-level event is linked to a psychological event that connects the psychological to the quantum. Each human conscious experience is seen as a "feel" of an event in the top level of processing in the human brain. Stapp sees the physical world as a structure of tendencies or probabilities within the world of the mind.

Von Neumann introduced an ontological approach that brings the observer and the instrument in the state of the system. Stapp describes Von Neumann's view of the quantum theory defined as "state of the universe is an objective compendium of subjective knowings." This statement describes the fact that the state of the universe is represented by a wave function that is a compendium of all the wave functions that all of us can cause to collapse with our observations.

Stapp mentions the Quantum Zeno effect, which is a phenomenon in which a system is "freezed" if we keep observing the same observable rapidly. We have to make a conscious decision about which question to ask nature (which is observable to observe?). The rapid sequence of process 1 actions in von Neumann theory, triggers a quantum Zeno effect, which holds the brain state in an associated subspace much longer than what process 2 would allow. This promotes the causal efficacy of conscious will.

Beck and Eccles Approach

The information flow between neurons in chemical synapses is initiated by the release of transmitters in the presynaptic terminal. This process is called exocytosis and is triggered by an arriving nerve pulse with some small probability. Beck and Eccles have shown that quantum processes are distinguishable from thermal processes for energies higher than 10^{-2} eV (at room temperature). For typical length scale for biological microsites of the order of several nanometers, an effective mass below ten electron masses is sufficient to ensure that quantum processes prevail over thermal processes.

The upper time scale limit of such quantum processes is of the order of 10^{-12} sec. This is significantly shorter than the time scale of cellular processes, which is 10^{-9} sec and longer. The significant difference between the two-time scales makes the two processes decoupled from one another.

Beck and Eccles proposed trigger mechanism as tunneling processes of two-state quasiparticles resulting in state collapses. It yields a probability of exocytosis in the range between 0 and 0.7, in agreement with empirical observations. However, the question remains how the trigger may be relevant to conscious mental states. There are two aspects to this question.

The first aspect refers to mental powers on brain states. This is conceived in a way that mental intention becomes neurally effective by momentarily increasing the probability of exocytosis. The second aspect refers to the fact that processes at synapses cannot be simply correlated to mental activity, whose neural correlates are coherent neural assemblies. This remains an unsolved problem. The approach by Beck and Eccles essentially focuses on brain states and brain dynamics.

Karl Pribram Approach

The holonomic brain theory, developed by neuroscientist Karl Pribram initially, in collaboration with physicist David Bohm, is a model of human cognition that describes the brain as a holographic storage network. Many properties of the brain are the same properties as associated with holograms. Memory is distributed in the brain, and memories do not disappear all of a sudden but slowly fade away. Pribram suggests that these processes involve electric oscillations in the brain's fine-fibered dendritic webs, which are different from the more commonly known action potentials involving axons and synapses. These oscillations are waves and create wave interference patterns in which memory is encoded naturally, and the waves may be analyzed by a Fourier transform.

In Pribram's opinion, a sensory perception is transformed in a brain wave, a pattern of electromagnetic activation that propagates through the brain just like a wave front in a liquid. This activity in the brain creates the interpretation of the sensory perception in the form of a "memory wave," which in turn crosses the brain. The interference of a memory wave and a perceptual (e.g., visual) wave generates a structure that resembles a hologram.

Pribram employs Fourier transforms to deal with dualism between space-time and spectrum. All perceptions can be analyzed into their component frequencies of oscillations and therefore treated by Fourier analysis. Gabor's uncertainty principle, that states that exact time and frequency of a signal cannot be known simultaneously, sets a limit at which both frequency and space-time can be concurrently determined. The fundamental minimum is Gabor quantum of information.

Pribram suggests that consciousness may occur primarily in the dendritic-dendritic processing and the axonal firing may support primarily automatic nonconscious activities.

My Approach

Umezawa and Takahashi described an interesting physical process for memory retrieval in the brain in terms of quantum field theory. As long as memory is maintained in the geometric objects of the quantum electric dipole field, Nambu-Goldstone bosons emerge from the geometric objects due to breakage of symmetry triggered by arbitrarily small incoming energy. Emergence of Nambu-Goldstone bosons is memory retrieval. Then what is consciousness? It should be some kind of physical property created by the Nambu-Goldstone bosons. I proposed that the physical property can be explained by Yukawa coupling, named after Hideki Yukawa, which is an interaction between a scalar field f and a Dirac field φ. Nambu-Goldstone bosons create the scalar field and the Dirac field is created by fermions like electrons. The Yukawa coupling is given by,

$$V \approx K \varphi f \varphi^{\prime}$$

Where K is the Yukawa coupling factor and V is the energy transfer due to Yukawa interaction.

The Dirac field for an electron is:

$\varphi = m_e e^{jwt}$, $\varphi^{\prime} = m_e e^{-jwt}$ where m is the electron mass.

Hence, $V \approx K f m_e^2$

The wave function of the f of the Nambu-Goldstone boson is given by

$$f = \rho(x) e^{j\theta(x)}$$

where $\rho(x)$ is the local density of the condensate and $\theta(x)$ is the phase.

Hence $V(x) \approx K m_e^2 \rho(x) e^{j\theta(x)}$

The electrons causing the Yukawa coupling come from the axons of the various neurons in the brain for a certain event. Hence the energy *V(x)* from each coupling will have different phases. If we call these energy levels *V(x1), V(x2), V(x3), etc.,* then these energy levels will form an interference pattern creating an image which is the consciousness created in the brain.

The image of consciousness created by the interference pattern is stored in a certain location in the brain. This could be similar to microtubular substructure of the neuron as proposed by Hameroff and Penrose. Neurons are filled with the intricate structure of microtubules. Each microtubule is a cylindrical structure many millimeters long. A spiral chain of tubular molecules forms the surface of the cylinder. Each tubulin molecule has a single special electron that can be in one of two stable locations. Each stable location should be able to store a bit "0" or "1." Hence, each location of the tubulin molecule can hold a "bit" of information, as with computer memory. This may be how the image of consciousness is stored digitally in the microtubular substructures of the neurons in the brain-like computer memory.

Consciousness obtains energy from zero-point energy to interpret the interference pattern of the energy levels generated by the Yukawa coupling of scalar fields of Nambu-Goldstone bosons and the Dirac field of electrons before it collapses. Zero point is described in appendix 1. Consciousness merges with the supreme consciousness, which is the zero-point energy of super universe.

Conclusion

No one fully understands quantum mechanics. Schrödinger said, "Had I known that we were not going to get rid of this damned quantum jumping, I never would have involved myself in this business." This shows the frustration of one of the founders of quantum mechanics with the inexplicability of quantum mechanics. Physicist John Wheeler said, "If you are not completely confused by quantum mechanics, then you don't understand it." But quantum mechanics has been extraordinarily successful in explaining the behavior of atomic and subatomic particles, which classical mechanics could not.

Consciousness is perhaps one of the most controversial areas of research in neurology and philosophy. Currently, there is no consensus on how to define, explain, or measure consciousness. The human brain is one of the most complex systems we know. It has around one hundred billion neurons communicating in trillions of connections called synapses. Since classical physics could not explain consciousness, it is quite natural to approach it from quantum mechanics. Quantum theory has been intriguing for scientists to provide a physical explanation of consciousness. There can be no reasonable doubt that quantum events occur and are as efficacious in the brain as elsewhere in the world, including the biological systems. But it is still not resolved that these quantum events are related to consciousness. Many scientists, as discussed in this chapter, have proposed the quantum model of consciousness. They are speculative, with varying degrees of elaboration and viability. However, there is no consensus among them. Collapse type of quantum events introduce an element of randomness that is ontic rather than epistemic.

There is no doubt that scientists' vigorous pursuit to explain consciousness with quantum theory will continue. In the next chapter, we will discuss consciousness and cosmology.

CHAPTER 6

Cosmic Consciousness

In this chapter, we will discuss cosmology and its relation with consciousness. Cosmology is the science of the origin and development of the universe. Cosmology has been the subject of interest since early civilization. When people looked at the sky and saw the moon, twinkling stars, the morning and evening star, they wondered about the vast space and sky, our place in it, and its future. Cosmology is studied by scientists, such as astronomers and physicists, as well as philosophers, such as metaphysicians, philosophers of physics, and philosophers of space and time. Because of this shared scope with philosophy, theories in physical cosmology may include both scientific and nonscientific propositions—and may depend upon assumptions that cannot be tested. Cosmology differs from astronomy in that the former is concerned with the universe as a whole, while the latter deals with individual celestial objects.

Origin of the Universe

Before the big bang theory, steady state theory was prominent in explaining the origin of the universe. According to steady state theory, the universe has no beginning and no end. Because the universe is expanding, new matters are being created continually to keep the density from decreasing. However, the discovery of cosmic microwave background (CMB) radiation by Arno Penzias and Robert Wilson in 1965 refuted the credibility of steady state theory. They found that CMB bathes the earth evenly in all directions.

This cannot be explained by the scattering of radiation by galactic dust particles as proposed in steady state theory. This is due to cooling of the universe as it expands, as explained by the big bang theory. Since the discovery of CMB, the big bang theory is considered the best theory of the origin of universe.

Big Bang Theory

According to the big bang theory, the universe was created from a point of infinite energy called singularity approximately 13.7 billion years ago. After creation of singularity, between time 10^{-10} s < t <1 s at temperature 10^{15} K > T > 10^{10} K, free electrons, quarks, neutrons, photons and neutrinos were strongly interacting with everything else. Between 1 s < t < 10^{12} s at 10^{10} K > T > 10000 K, protons and neutrons joined to form atomic nuclei, and we have free electrons, quarks, neutrons, photons, and neutrinos; everything is strongly interacting with everything else except the neutrinos, whose interactions are too weak. The universe is still radiation dominated. Between 10^{12} s and the present age at 10000 K > T > 3 K, atoms formed from the nuclei and electrons since photons were no longer interacting with them and cooled to form microwave background.

Observational evidence for the big bang comes from the redshift of the spectra of galaxies proportional to their distance, as described by Hubble's law. Redshift is the Doppler effect applied to light waves. Doppler effect states that the frequency from a light wave of a moving object decreases as it moves away from us. Stars have sets of absorption and emission lines identifiable in their spectra. If a star is moving toward us, its light waves get crowded together, raising the frequency. Because the blue light is at the high-frequency end of the visible spectrum, this is known as blueshift. If the star is receding, the spectrum lines move toward the red end and the effect is known as redshift. American astronomer Edwin Hubble (1889–1953) discovered a proportionality of the distance of the stars with their redshift. Hubble and American astronomer Milton Humason were able to plot a trend line from the forty-six galaxies they studied and proposed Hubble's law, which states that the radial velocity, v, with which a galaxy recedes is linearly proportional to its distance, r from us and $v \propto r$; $v = H.r$, whereas H = Hubble constant. The current estimate of Hubble constant H from measurements using the Hubble Space Telescope is H = 72 kilometers per second per megaparsec (1 parsec = 3.261 light-years = 3.07×10^{13} kilometers; 1 megaparsec = 3.07×10^{19} kilometers).

$H = 2.35 \times 10^{-18} \text{ sec}^{-1}$

Now, the distance r is given by the velocity v multiplied by time t as follows: $r = v.t$. Since $v = H.r$, hence $t = 1 / H$.

Substituting the value of H, $t = 1 / (2.35 \times 10^{-18} \text{ sec}^{-1}) \cong 13.5$ billion years, which is close to the estimated age of the universe.

Although the big bang theory matches well with experimental results, the unsolved question is of the origin of singularity. I have proposed a new theory for the origin of singularity in the current big bang theory, based on zero-point energy. Our observable universe is a part of the entire universe, which is infinite. According to Planck's equation, there is energy even at absolute zero temperature as hf/2, where h is Planck's constant and f is frequency. According to my theory, zero-point energy exists in the entire universe, and I have mathematically shown that zero-point energy from zero to Planck frequency can combine to create energy of a colossal amount, similar to the singularity of the big bang theory. Zero-point energy has been experimentally tested and proven that it exists (appendix 1).

The nucleus of an atom consists of protons and neutrons. A proton has a positive charge, while a neutron has no charge. Protons and neutrons are made of quarks bonded by gluons. Leptons are electron, muon, and tau with negative charge. A muon is two hundred times heavier than an electron; a tau is heavier than a muon. Temperatures during this period were so high that the random motions of particles were at a significant fraction of the speed of light, and particle-antiparticle pairs of all kinds were being continuously created and destroyed in collisions.

The universe continued to grow in size and fall in temperature, hence the typical energy of each particle was decreasing. After about

10^{-11} second, the picture became less speculative since particle energies dropped to values that can be attained in particle physics experiments. The temperature was about 10^{15} K, and free electrons, quarks, photons, and neutrinos were strongly interacting with each other. One second after the initial explosion, the temperature dropped to 10^{10} K, photons no longer had the energy to disrupt the creation of nucleus consisting of neutron and proton, and nucleosynthesis began to form nucleus of an atom. The universe was still radiation dominated. At 10^{13} seconds after the initial explosion, when the universe was three hundred thousand years old, the temperature dropped to 3,000 K and atoms formed from nuclei and electrons. The photons were no longer interacting with them and were cooling to form what is known as the microwave background. This process is known as decoupling. The decoupling happened when the universe was one-thousandth of its present size. In the early universe, the only elements produced in any significant abundance were hydrogen and helium-4 (nucleus with two protons and two neutrons). Helium-4 was produced since it has the most stable light nucleus, and hydrogen was produced because there were not enough neutrons around for all protons to bind with; therefore, some protons were left over.

After decoupling, clouds of gas and dust, called nebulae, began to form, which were constantly in motion. As a result, some regions in the nebulae periodically had a higher concentration of gas and dust than others had, causing stronger force of gravity in those regions. When the force of gravity was sufficiently strong in a particular region, a star was formed. As the collection of gas and dust in the star continued, the temperature at the center rose higher and higher. When the temperature reached eighteen million degrees Fahrenheit, nuclear fusion took place and a star was born. Many stars are between one billion and ten billion years old. Some stars may even be close to 13.7 billion years old—the observed age of the universe. The more massive the star, the shorter its life span, primarily because massive stars have greater pressure on their cores, causing them to burn hydrogen more rapidly. The most massive stars last an average of about one million years, while stars of minimum mass (red dwarfs) burn their fuel slowly and last tens to hundreds of billions of years.

Composition of the Universe

The fundamental building blocks of our universe are the galaxies. Galaxies are in various types—some spiral, some elliptical, and some with irregular shapes. Our galaxy is the Milky Way. Galaxies come with a wide range of masses. Some galaxies have only a million solar masses and some could have ten times more than the Milky Way. Some galaxies group together to form galaxy clusters. The number of galaxies in the observable universe range from two hundred billion (2×10^{11}) to two trillion (2×10^{12}) or more. Stars form the galaxies.

Galaxies have supermassive black holes in the center. A black hole is a super-condensed mass whose gravitational force is so high that even light cannot escape. The Milky Way galaxy, home of Earth and the solar system, appears to harbor at least one such black hole within its nucleus. The Milky Way galaxy system has at least two hundred billion other stars (more recent estimates have given numbers around four hundred billion) and their planets, in addition to thousands of clusters and nebulae. All the objects in the Milky Way galaxy orbit their common center of mass, called the galactic center. As a galaxy, the Milky Way is actually a giant, as its mass is probably between 750 billion and one trillion solar masses.

Our star sun, together with the whole solar system, is orbiting the galactic center on a nearly circular orbit. We are moving at about 156 miles per second and need about 220 million years to complete one orbit. Therefore, the solar system has orbited the galactic center about twenty or twenty-one times since its formation about 4.6 billion years ago. The Milky Way consists of the flowing areas:

1. The galactic center and bulge around the center
2. The disk with spiral arms, which contains the majority of the stars, including the sun, and virtually all of the gas and dust
3. The halo, a roughly spherical distribution, which contains the oldest stars in the galaxy

The galactic center harbors a compact object of very large mass (named Sagittarius A), strongly suspected to be a supermassive black hole. Most galaxies are believed to have a supermassive black hole at their center. The bulge is composed primarily of red stars and molecular hydrogen gas. The galactic disk has a diameter of between seventy thousand and one hundred thousand light-years. The disc is believed to have four major spiral arms that all start at the galaxy's center. The sun is located in one of the spiral arm about two-thirds of the way from the center to the edge of the disk (about twenty-five thousand light-years by the most modern estimates). The galactic disk is surrounded by a spheroid halo of old stars and globular clusters, with a stellar halo diameter of two hundred thousand light-years. However, a few globular clusters have been found farther away, at more than two hundred thousand light-years from the galactic center.

Our star sun has enormous bulk and mass. The pressure at the center of sun is extremely high, as is the density (150 times the density of water on Earth), and the temperature is around fifteen million degrees). The sun maintains its stable equilibrium under the opposing action of two gigantic forces: the gravitational force pulling inward and the pressure of gas pushing outward; these exactly balance each other. The sun has been shining for 4.6 billion years and has so far converted about one hundred Earth masses into energy. At this rate, it has enough energy to last another five billion years.

The most recent Wilkinson Microwave Anisotropy Probe (WMAP) observations predict that the universe is made of 74 percent dark energy, 22 percent dark matter, and 4 percent baryonic matter (hydrogen, helium, stars, neutrinos, and others), as shown in figure 13. It has been found that the mass of the baryonic matter in the galaxies is not enough to explain the rotation of the galaxies. The centrifugal acceleration and the gravitational pull, due to mass in the galaxy, must balance out for the galaxy to be stable. The visible part of the galaxy that contains the baryonic matters is within radius R. The mass outside the radius R does not contribute to the gravitational pull. Hence, the rotational velocity should drop off as the square root of radius R. Instead, it is more or less constant. This is due to dark matter, which is nonbaryonic and interacts extremely weakly with conventional matter.

Dark energy solves the missing mass problem of the universe. For the universe to be flat, the mass/energy density of the universe must be equal to a certain critical density, 10^{-26} m^{-3}. Astronomers have now found that the universe is not only expanding but also expanding at an accelerating rate. Since matter cannot expand by itself, the accelerated expansion must be due to dark energy.

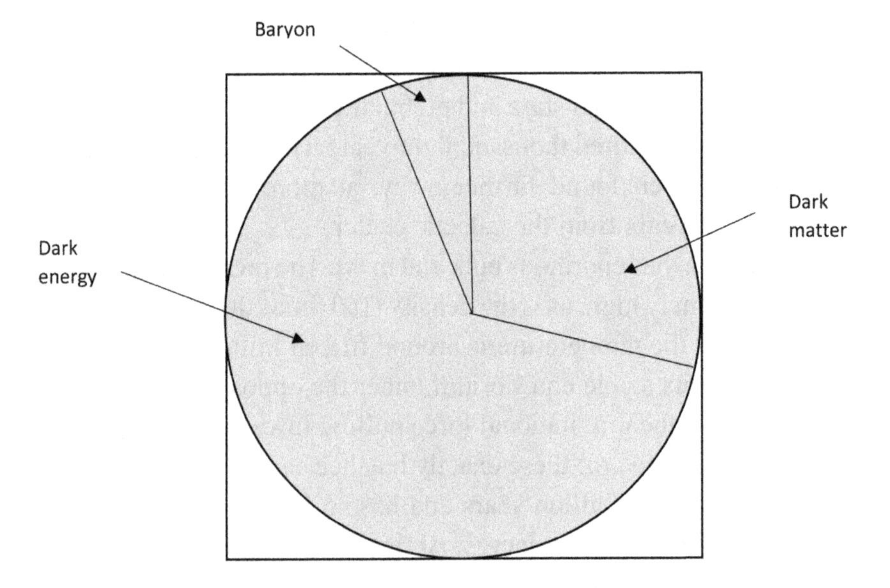

Figure 13: Composition of the universe

Modern physical cosmology is dominated by the big bang theory, which attempts to bring together observational astronomy and particle physics.

Philosophy and the Universe

Prakriti and purusha in Hinduism participate, regulate, and implement the universal creative process. They are the two different aspects of the manifest Brahman. Prakriti means that which is found in its natural, unaltered form. Its opposite is vikriti, which means that which is deformed or altered from its natural state. Prakriti also means "that which gives shapes or forms," signifying nature or pure energy.

Purusha means manifested Brahman or the creative consciousness that sets in motion the entire creative process. Both purusha and prakriti are eternal, indestructible realities.

> Physicists are being forced to admit that the universe is a "mental" construction. Pioneering physicist Sir James Jeans wrote the following: "The stream of knowledge is heading toward a nonmechanical reality; the universe begins to look more like a great thought than like a great machine. Mind no longer appears to be an accidental intruder into the realm of matter; we ought rather hail it as the creator and governor of the realm of matter. Get over it and accept the inarguable conclusion. The universe is immaterial-mental and spiritual.
>
> —R. C. Henry, professor of physics and astronomy at Johns Hopkins University, "The Mental Universe," *Nature* 436, no. 29 (2005)

Some philosophers have suggested that consciousness is a fundamental component of the universe. As we discover new phenomenon in the universe that cannot be explained by existing laws and theories, scientists propose new fundamental laws. When classical physics could not explain the structure of the atom, quantum mechanics was developed. Existing fundamentals such as space, time, and mass could not explain electromagnetic phenomena, so Maxwell developed electromagnetic theory. For thousands of years, consciousness could not be explained, so philosopher David Chalmers suggested that consciousness is a fundamental building block of nature. He also suggested that every system down to the very elementary particles have some kind of consciousness.

Universal consciousness is a concept that tries to address the underlying essence of all being and becoming in the universe. It includes the being and becoming that occurred in the universe prior to the arising of the concept of "mind," a term that more appropriately refers to the organic human aspect of universal consciousness. It addresses inorganic being and becoming and the interactions that occur in that process without specific reference to the physical and chemical laws that try to describe those interactions. Those interactions have occurred, do occur, and continue to occur. Universal consciousness is the source, ground, and basis that underlies those interactions and the awareness and knowledge they imply.

According to Deepak Chopra, cosmic consciousness, it isn't just real—it's totally necessary. It rescues physics and science in general from a dead end—the total inability to create mind out of matter—and gives it a fresh avenue of investigation. The Higgs boson has gotten us a bit closer to a unified field theory—only a bit—but we are still far away from a full theory of quantum gravity. In many versions of superstring theories, the so-called M-theories, it is deduced that a vast number of parallel universes exist, all forming what is called the multiverse. However, the multiverse cannot be an explanation of why this particular universe of ours is what it is. Having a vast number of universes emerging from empty space still does not explain why consciousness is in our universe. The entire universe is a matter of transformation whereby something is available to be turned into perception. Consciousness is that something.

From a singularity, out of nothingness emerged the universe, which emerged upon becoming conscious of itself. By becoming conscious, this caused a "collapse of the wave function" and the universe underwent a transition "from the possible to the actual." Thus if the universe became conscious of itself, it collapsed the wave function and the universe was transformed into the actual. Since the universe, as a collective, must have a wave function, then this universal wave function would describe all the possible states of the universe and thus all possible universes. Hence there must be multiple universes that exist simultaneously as probabilities.

Panpsychism

A new theory called panpsychism is gaining ground in scientific circles. In this theory, the entire universe is inhabited by consciousness. Panpsychism sounds very much like what the Hindus and Buddhists call the Brahman, the tremendous universal Godhead of which we are all a part. In Buddhism, for instance, consciousness is the only thing that exists. Panpsychism can also be seen in ancient philosophies such as Stoicism and Taoism. The word *panpsychism* was coined by the Italian philosopher Francesco Patrizi in the sixteenth century and derives from the two Greek words *pan* (all) and *psyche* (soul or mind). Baruch Spinoza and Gottfried Leibniz can be said to be panpsychists in the seventeenth century. Spinoza considered "God, or nature," as the one single infinite and eternal substance. Leibniz considered an infinite number of simple mental substances called monads, which make up the fundamental structure of the universe. Panpsychism was at its peak in the nineteenth century, and many philosophers and psychologists promoted panpsychism. Alfred North Whitehead was the most significant proponent of panpsychism in the twentieth century.

Contemporary panpsychism has taken the following two possible forms:

> *Panexperientialism*: the view that conscious experience is fundamental and ubiquitous
>
> *Pancognitivism*: the view that thought is fundamental and ubiquitous

Panexperientialism holds that this diminishing of the complexity of experience continues down through plants and through to the basic constituents of reality, perhaps electrons and quarks. Thought, in pancognitivism, is a much more sophisticated phenomenon, and many doubt that it is correct to ascribe it to nonhuman animals, never mind fundamental particles. Panexperientialist forms of panpsychism are taken seriously in contemporary analytic philosophy.

David Chalmers distinguishes between constitutive and nonconstitutive forms of panpsychism, a distinction in the following way:

> Constitutive panpsychism: Forms of panpsychism according to which facts about human and animal consciousness are not fundamental but are grounded in/realized by/constituted of facts about more fundamental

> kinds of consciousness—for example, facts about microlevel consciousness.
>
> Nonconstitutive panpsychism: Forms of panpsychism according to which facts about human and animal consciousness are among the fundamental facts.

According to constitutive *micropsychism*, the smallest parts of the brain have basic forms of consciousness, and the consciousness of the brain as a whole is, in some sense, made up from the consciousness of its parts.

In his 1927 book *The Analysis of Matter*, Bertrand Russell proposed a novel approach to the mind-body problem that has become known as "Russellian monism." This can be defined in terms of two components, one negative and one positive

In some significant sense, the information we get from the physical sciences is limited. There are subtle variations on how exactly this is put, but the idea is that the physical sciences only tell us about the extrinsic, relational, mathematical, or dispositional nature of matter and leave us in the dark about its intrinsic, concrete, and categorical nature. Physics tells us how an electron behaves, but it doesn't tell us how it is in and of itself.

The intrinsic/concrete/categorical features of matter that physical science remains silent on account for the existence of consciousness. The problem of consciousness, the difficulty seeing how consciousness fits into the physical word, is the result of our not taking into account these "hidden" features of the physical world.

Russellian monism assumes that conscious states just are the intrinsic nature of brain states—hence the causal action of brain states and the causal action of conscious states are arguably the same thing. Thus it avoids both the deep problems facing dualism and physicalism. Russellian monism's elegant integration of consciousness in the material world looks to be a promising way of accounting for the causal role of human consciousness. If conscious states just are the intrinsic nature of brain states, then the causal action of brain states and the causal action of conscious states are arguably the same thing. The growing prominence of Russellian monism, given that one paradigmatic form of Russellian monism is panpsychism, has resulted in panpsychism being considered as a serious option.

The panpsychist explains human and animal consciousness in terms of more basic forms of consciousness. These more basic forms of consciousness are then postulated as properties of the fundamental constituents of the material world, perhaps of quarks and electrons. Thus, we try to explain the consciousness of the human brain in terms of the consciousness of its most fundamental parts.

Consciousness as fundamental force

The universe has four fundamental forces:

> Gravity: It works between two masses. Gravitational force is the product of gravitational constant, product of two masses, and inverse of the square of distances between the two masses.
>
> Electromagnetic: It acts between charged particles and is the combination of all electrical and magnetic forces. The electromagnetic force can be attractive or repulsive.
>
> Strong force: It binds protons and neutrons in the atomic nucleus and hold quarks in elementary particles like protons and neutrons.
>
> Weak force: It causes the beta decay (the conversion of a neutron to a proton, an electron, and an antineutrino).

Consciousness is another fundamental force of the universe. It works in any living object to a state being characterized by sensation, emotion, volition, and thought.

Conclusion

Big bang theory is the current accepted theory for the creation of our universe. According to this theory, our universe was created about 14.7 billion years ago from a point of infinite energy called singularity. Although the source of singularity is not known, the author has proposed a theory that singularity came from zero-point energy. I have also proposed that other parallel universes are created from zero-point energy. The discovery of the cosmic microwave background radiation in 1964, which was predicted by the big bang theory, has confirmed more than the big bang theory. Our universe is expanding at an accelerated rate as confirmed by astronomer Hubble by measuring the redshift of radiation and analyzing by Doppler shift. The diameter of the observable universe is estimated to be about 28.5 gigaparsecs (93 billion light-years, 8.8×10^{23} kilometers or 5.5×10^{23} miles).

The number of galaxies in the observable universe range from two hundred billion (2×10^{11}) to two trillion (2×10^{12}) or more. Our galaxy is the Milky Way, and our star sun is one of the one hundred billion stars in the Milky Way. Our universe consists of 74 percent dark energy, 22 percent dark matter, and 4 percent baryonic matter. Dark energy, although the source is unknown, provides the pressure for the universe to expand. Consciousness is also a fundamental building block of the universe. Like space, time, mass, and charge, consciousness also affects our existence in the universe. More and more philosophers and scientists are considering exploring this aspect.

CHAPTER 7

Time and Consciousness

Physical events happening outside and the same events reflected in our minds march along exactly in step, with the stream of actual moments in the outside world and the stream of conscious awareness of them synchronized perfectly. The cinema industry depends on the phenomenon that what seems to us as a movie is really a succession of still pictures running at twenty-five frames per second. We don't notice the break because the "now" of our conscious awareness stretches for over 1/25 second. Different levels of consciousness may experience time in quite different ways. This is evidently the case in response time. We jump at the sound of a telephone a moment or two before we actually hear it ring. The shrill noise induces a reflex response much faster through the nervous system than it takes to create the conscious experience of the sound.

Speech ability mostly belongs to the left side of the brain, whereas the primary auditory cortex on the right side of brain is involved in music appreciation. However, why should both experience a common time? Why should the subconscious use the same mental clock as the conscious? Sensory deprivation can also alter the impression of time intervals. Yogis practicing meditation claim that they can more or less suspend the flow of time altogether by detaching themselves from worldly events.

Phi Phenomenon

The phi phenomenon is the optical illusion of perceiving a series of still images, when viewed in rapid succession, as continuous motion. The classic phi phenomenon experiment involves a viewer or audience watching a screen upon which two small spots are lit briefly in quick succession at slightly separated locations. Typically, the spots are illuminated for 150 milliseconds, separated by an interval of a fifty-millisecond gap. The viewer reports seeing not a succession of spots but a single spot moving continuously back and forth. Evidently, the brain fills in the fifty-millisecond gap. This hallucination occurs after the event because until the second light flashes, the viewer cannot know the light is supposed to move. This hints that the viewer is not simultaneous with the visual stimulus but a bit delayed, allowing time for the brain to reconstruct a plausible fiction of what happened a few milliseconds before.

In further experiments, the first spot is colored red, the second spot green. The viewer reports seeing the spot change from red to green color abruptly in the middle of the trajectory and is able to indicate exactly where using a pointer. This result leaves us wondering how viewers can apparently experience the green color sensation before the green spot lights up. Is this consciousness projected backward in time? In another experiment, the subject wears a device that delivers light taps to the arm in certain sequence: a few to the wrist, followed by a couple to the elbow, then the shoulder, in rapid succession. When this is done, the subject reports the sensation of equidistant taps traveling up the arm, like a little animal hopping. It means that some taps are felt in between the points of contact, such as on the forearm. Here we have the mystery of how the brain knows it is going to receive an elbow tap after the wrist tap in order to create the false impression of a forearm tap in between. Is this a case of backward consciousness?

In another experiment, Benjamin Libet of the University of California attached electrodes to the exposed brain of a patient during a brain operation. By stimulating electrically, he was able to stimulate tingle in the patient's hand. Libet also attached electrodes to the skin of the hands. Therefore, he was able to compare the experiences of tingles reported by the patient when the hands and the cortex were both stimulated. In the first part of the experiment, Libet found that the actual sensation of the tingle occurred up to half a second after the stimulus was delivered to either the hand or the cortex, even though the signal's travel time to the brain was only about ten milliseconds. In the second part of the experiment, Libet stimulated the left hand at the same as the left cortex. The latter produced a tingle in the right hand, so the patient felt tingles in both hands and reported which one seemed to occur first. It is general to assume that since the cortex is closer to the "seat of consciousness" than the hand, the brain-induced tingle in the right hand would be experienced before the skin-induced tingle in the left hand. But the time order was completely reversed. The patient definitely felt the left hand tingle first. Even when the hand was stimulated a short while after the brain, the order was reversed. Libet explained his unexpected results by claiming that when the skin is stimulated, the sensation is referred back in time to when it actually occurred, whereas no such backward referral takes place for cortical simulation.

Electrodes attached to the scalp can monitor brain waves and detect bursts of activity that are associated with voluntary movements, such as flexing a finger. With his research team, H. H. Kornhuber found that in some cases the brain cells start working as long as a second or more before the physical movement actually begins. It is as if the brain knows what you are going to do moments before you decide to do it. This precursory electrical burst has been called "the readiness potential" by philosopher Karl Popper and neurophysiologist Sir John Eccles.

Consciousness of time differs in a significant respect from the consciousness of other physical qualities such as size or shape. When we see a square, the electrical activity in our brains is not square shaped. There is no little square inside our heads projected on a movie screen for us to watch. Instead, a complicated pattern of electrical activity somehow produced the sensation of square. Hence, square is represented by an electrical pattern, but representation is not the same as the pattern of the object. When it comes to time, the situation is more complicated.

Time and Brain

Neuroscientists found that lateral intraparietal cortex (LIP), which is in the lateral surface of the parietal lobe, plays a key role in keeping track of time. Firing patterns of neurons in LIP keep track of time. To some extent, all cognitive functions in the brain rely on time. Hence, timing is just not limited in LIP but occurs throughout the circuits of the brain. According to Professor Mehrdad Jazayeri of MIT, firing patterns of neurons change with time of an action. When an action like reading a sentence is completed at different speeds, the timing of firing patterns of neurons changes accordingly. When the sentence has to be read quickly, the firing speed of the neurons is faster. When the sentence is read slowly, the firing speed of the neurons is slower. Hence, there is no central clock in the brain to keep time.

We use time and distance interchangeably. If someone asks how far London is from New York, we normally say seven hours by flight. But if we ask how far Boston is from New York, the answer we get is 215 miles. Therefore, space and time are interchangeable depending on the situation. So does it mean that neurons may be functioning in single time-space dimension?

Our brains accumulate evidence at a slower speed. Information about the present conscious event is outdated by at least one-third of a second. It could be even slower if the information is so vague that it needs a slow accumulation of evidence before it crosses the threshold of conscious perception. This is similar to an astronomer accumulating faint lights from a distant star by long exposure on a photographic plate. However, we are unaware of this delay. We do not realize that our subjective perception lags behind the objective events in the outside world. The color we see and the sound we hear date from at least one-third of a second ago. This could be due to the time taken by sensual reception by our sense organs, followed by time taken for neural processing in our brains.

Anticipation is a mechanism that compensates for the sluggishness of our consciousness. Our sensory and motor areas in the cortex of the brain contain temporal learning mechanisms that anticipate events in the outside world. This lets us perceive events closer to the time when they actually occur. When our anticipation mechanism fails, we become aware of the delay that our consciousness imposes. When a flash of light appears, we do not instantly turn away from it. We first are attracted to it and only later turn away from it.

Subjective and Objective time

Objective time is the time measured by a clock and can be verified. Subjective time is based on personal view. One hour measured by a clock may be long for one person but short for another person. Analytic means to break down, and distinguish is the essence of the modernization process. We can even get so analytic that we can never fully fuse with the world in selfless activity to be fully engaged. When we create something, it does not mean it is not real. Indeed, the societies that human beings create are real. Their reality is simply different from nature. We present our measured time as real time and call it objective. But the sense of time—the sense of duration—is dependent on our mood or interpersonal differences. We call it subjective and regulate it to a less important, less real realm. From the nonpartisan viewpoint of gaining systematic knowledge, we will benefit from differentiating between the subjective reality of time as privately experienced and the objective reality of time that is shared and public. We may experience a fifty-minute class time differently from our professor—that is, subjectivity. But without objective time, we wouldn't be able to know we had different experiences of the same (public) time. Objective knowledge does not draw us close to nature. Objectivity removes us from the possibility of mirroring the world or fusing with the world. Our minds separate us from it. The natural function of human mentality seems to control by detachment, not to fuse to oneness.

Cognition developed out of the Greek word for thinking; to cogitate is to think. A crucial characteristic of thought or idea is that they are not so locatable in time and space. Cognition is not so locatable. Let us return to the notion of "space." As enlightenment thinker Immanuel Kant understood, when I look out of my window I do not perceive "space." I see grass and an embarrassing amount of leaves. Down the hill, I see trees and my cat. All these objects are substantial and locatable; they have color, smell, and even tastes. They have their own feel. But where is space in all of this? What color is it? What does it smell like? Every tangible thing has some kind of smell. How does space feel when you touch it? Indeed, can you touch it? Are these appropriate questions to ask in the first place? Kant knew they were not. Space is of a different order than the perception of substantial things. It is a way of putting organization on all these stimuli, but it is not of the same order of these stimuli. Space is a concept, a cognition, not a sensation.

The same thing goes for time. We grasp it with our minds, not our hands. We give it material form in various ways, like the dots to which hands on round clocks point, or digital watches that show only numbers, or even sundials. There are countless ways to portray time in visible form, but these are perceptual vehicles for conveying extra sensory time. These are not time itself. Time per se can only be grasped cognitively. The different sensed expressions of time are different ways of communicating the same extrasensory, intangible principal. Again, if we are to understand the cultural variability in emotions, we must understand the cultural variability in the thought that makes up these emotions. As with all cognition, we should not be surprised at all to find that many societies have quite different notions of time. In contrast to our sensory windows to the world, whose origins are in material nature, the origin of cognition is at large in society and particular languages. Cognition is culturally variable. If cognition is a part of emotion, then many emotions can differ from culture to culture. Of course, it appears that common sense is just common sense—everywhere the same—but some cultures see the observable fact that night follows day as proof that time is a continuous line; others see it that time is forever cyclical—that is, repeating its self.

According to Einstein, the faster one goes, the slower time passes. If one travels in a spaceship nearly as fast as the speed of light, she might come back to Earth to find her friends much older than she is, despite the fact that they were her age when she left. These are two objective times in two different situations. Under Einstein's theory of general relativity, gravity can bend time. But what about subjective time?

Chronesthesia

Chronesthesia is mental time travel, when we think of the past, present, and future. This form of consciousness allows people to travel in subjective time. When I think of the vacation in Hawaii with my wife, I mentally travel in time at the speed I want, remembering all the wonderful things we did. This is totally subjective time, the speed of which changes continually depending on the situation. This is chronesthesia. Similar things happen when we think about the future. However, what is happening at present is objective time because we cannot change the measurement of time. It is the time a clock reads.

Scientists have mapped regions of the brain that could possibly be involved in mental time travel. They found that that the hippocampus and visuospatial cortex are involved in chronesthesia. The hippocampus is a small organ located within the brain's medial temporal lobe. Visuospatial cortex is located in the parietal lobe of the brain.

Conscious Experience of Time

Consciousness of time differs from other physical qualities such as spatial size or shape in a significant way. Our brains receive a desynchronized jumble of signals from which a consistent impression of time is built. It may be that the electrical patterns in the brain that represent time sequences may be quite different from the actual time sequence of the events they represent. Stuart Albert of the University of Pennsylvania did an interesting experiment. He shut volunteers in a room in which the wall clock had craftily been adjusted to run at either twice the speed or half the speed, without informing the subjects. Amazingly, their mental functions automatically adjusted to the accelerated or retarded pace. For example, when memory was tested, it was found to decay faster for subjects in the speeded-up group than the slowed-down group. Though our mental physiological functions are regulated by our neurological and chemical clocks buried within us, these clocks are not aware of the passage of time, an essential ingredient of pace judgment and sense of rhythm.

The question is whether subjective time or human psychological time has any relevance in Newton's or Einstein's objective time. There should be an aspect of time of greater significance that we have overlooked so far. According to physicist Arthur Eddington, we experience time in two distinct ways. The first is through our senses, in the same way we perceive our spatial relationships. The second way is to feel directly and deeply within our souls. Thus there is a sense of time buried deep within human consciousness, intimately associated with our sense of personal identity. In this way, Einstein's time or Newton's time makes no place for personal experience or sense.

Two areas might look at things differently; one is chaos theory, and the other is quantum physics. A chaotic system, although in a strict mathematical system deterministic, is highly sensitized to minute disturbances. Hence, meaning prediction is precluded. Chaos theory suggests that many physical systems are chaotic, but some, like the human brain, operate at the edge of chaos. Quantum physics also has a role here. Chaos in classical systems can amplify quantum fluctuations due to sensitivity to the smallest changes in initial conditions. The idea is that chaos could amplify quantum events, causing a single neuron to fire that would not have fired otherwise. If the brain (a macroscopic object) is also in a chaotic dynamical state, making it sensitive to small disturbances, this additional neural firing, small as it is, would then be further amplified to the point where the brain states would evolve differently than if the neuron had not fired. In turn, these altered neural firings and brain states would carry forward such quantum effects, affecting the outcomes of human choices.

Roger Penrose, John Eccles, and others have claimed that some cerebral processes are irreducibly quantum mechanical in nature. Attempts to explain the flow of time using physics are the most exciting contemporary developments in the study of time. Until we have firm understanding of the flow of time and its relation with consciousness, we will not know what part we are playing in the great cosmic drama.

Conclusion

Consciousness acts differently with time than shape or size. Different levels of consciousness may experience time in quite different ways. Phi phenomenon shows how consciousness is projected backward to fill the time gaps between two events to make it seem continuous. Neuroscientists have found that two brain regions—the hippocampus and the entorhinal—keep track of both space and time. Physics deals with objective time, which is precious and based on precise laws, whereas subjective time depends on individuals and their consciousness. The same objective time precisely measured by clocks will vary with individuals depending on their situation. Chronesthesia, by definition, is a form of consciousness that allows people to think about past and future and mentally travel in it.

In appropriating time with precise mathematical formula, physicists have robbed us of its original and human content. The value of human psychological time derives entirely from subjective factors and is unrelated to precise equations of physics. However, it would be wrong to discard entirely human experience of time as an illusion or misperception. Consciousness, time, and their interrelation have to be explored as a new branch of science with quantum physics and chaos theory.

In the next chapter, consciousness is explained as a completely different function.

CHAPTER 8

Consciousness as Physical Constant

Many theories have been proposed in explaining consciousness. Questions about the nature and origin of conscious awareness have likely been asked for as long as there have been humans. There have been two scientific views of consciousness: monistic and dualistic. Monistic view is that the brain does all the functions—as such, there is nothing called consciousness. Dualistic view is that brain and consciousness are two separate things. Monistic view considers the brain a supercomputer that can do all the necessary functions. Three recent dominant theories of consciousness are cognitive theories, neural theories, and quantum theories. However, many questions remain unanswered. If consciousness is related to brain, what is the relation … and where does consciousness stay? This has been a question for thousands of years.

The Human Brain

Let us now explain how the human brain works. The human brain is the center of the human central nervous system, located within the head. The adult human brain weighs on average about 1.2–1.4 kilograms (2.6–3.1 pounds), or about 2 percent of total body weight, with a volume of around 1,260 cubic centimeters in men and 1,130 cubic centimeters in women, although there is substantial individual variation.

Before we go into any further discussion, let us describe the function of the human brain. The dominant part of the brain is the cerebrum. The cortex is the outer layer of cerebrum. The cortex is about two to four millimeters thick and has cell bodies (neurons) and dendrites. The cortex is called gray matter. Underneath the cortex is the white matter, consisting of exclusively axons connecting neuron cells in the gray matter to each other.

The cortex has grooves called sulci (singular is sulcus) creating the characteristic folded appearance of the brain. There are three structural layers between the cortex and skull: pia, arachnoid, and dura.

> Pia: a thin membrane covering the surface of the cortex everywhere, including down the sulci
>
> Arachnoid: a thin, transparent membrane composed of fibrous tissue covering the pia
>
> Dura: a tough membrane above the arachnoid
>
> Skull: located above the dura

The cerebrum is a large part of the brain containing the cerebral cortex (of the two cerebral hemispheres), as well as several subcortical structures. This is called gray matter. The cerebrum is divided into left and right hemispheres, which are linked by a bridge of nerve fibers called the corpus collosum, which contains two hundred million fibers. The left hemisphere receives most inputs from and controls mostly the right side of the body. It is specialized for language, analytical skill, and reasoning. The right hemisphere controls mostly the left side of the body. It deals with visual pattern recognition and perception.

Figure 14 shows the different parts of the brain. The cerebral cortex is divided into four sections called "lobes": the frontal lobe, parietal lobe, occipital lobe, and temporal lobe, as shown in figure 14. Each lobe has a twin on the other half or hemisphere of the brain. The occipital lobes at the back farthest from the eyes largely process vision. The temporal lobes at the bottom middle of the brain process hearing and some aspects of language, especially in the left hemisphere, but are also where visual processing becomes object recognition. This is where our long-term memories are stored. The parietal lobes at the top back help process our sense of space as well as touch. The back portions of the parietal lobes are also linked with complex thought. The frontal lobes control important cognitive skills such as emotional expression, problem-solving, memory, language, judgement, and sexual behavior. In essence, it is the "control panel" of our personalities and our ability to communicate.

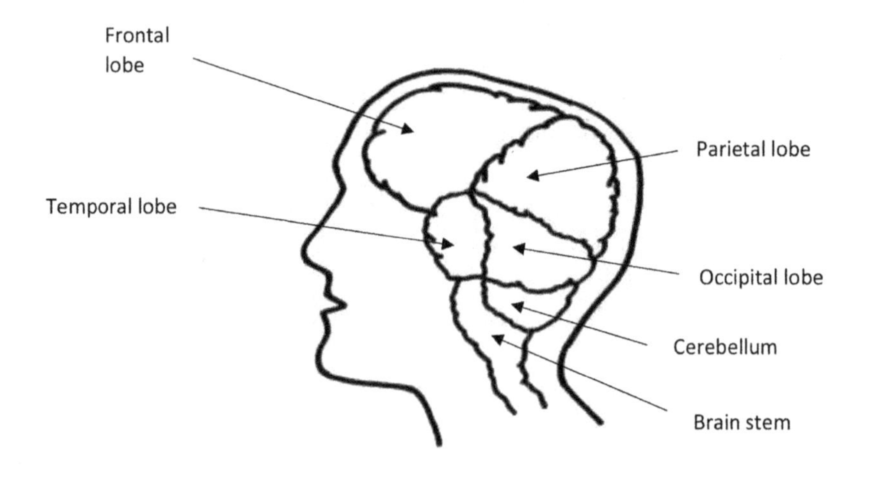

Figure 14: Lobes of cerebrum

The brain stem is the gateway between the brain and the body—all sensory signals from the body pass through the brain stem to the rest of the brain. In turn, all commands from the brain are shunted through the brain stem, down into the spinal cord, and through the rest of our bodies to make our motor commands seamlessly fulfilled. The brain stem consists of midbrain, pons, and the medulla.

> Midbrain: Serves important functions in motor movement, particularly movements of the eye and in visual and auditory processing.

Pons: Situated below the midbrain. Mediates auditory and balance functions.

Medulla: Situated below pons. The lower of the medulla is the tip of the spinal cord.

The most critical area for consciousness in the brain area is a part of the brain stem known as the reticular formation. It controls the sleep-wake cycle through a complex set of subregions, each playing a part in a chemical and neural set of activities.

The crucial parts of the brain structures are as follows:

Thalamus: The thalamus is a small structure within the brain located just above the brain stem, between the cerebral cortex and the midbrain. All signals from the senses are relayed through the thalamus. Its neurons receive inputs from, and send outputs to, almost every other brain region. It is a sophisticated information-filtering and information-organizing device and plays a central role in consciousness.

Cerebellum: The cerebellum is located between the occipital lobe and brain stem. It is responsible for coordinating motor behavior, balance, and posture. When we first learn to ride a bicycle, we have to think about everything we are doing. After learning, the process is programmed in the cerebellum and we don't think about each step.

Hippocampus: The hippocampus is located in the temporal lobe. It has crucial function in the creation of memories. It receives input from the entire cortex and projects back to the same areas.

Wernicke's and Broca's areas: Wernicke's area is located between the temporal lobe and parietal lobe. Its function is to process speech and language. Broca's area is located in the frontal lobe. It controls tongue, lips, and other speech articulators.

Amygdala: The amygdala is located in front of the hippocampus and is primarily involved in emotional processing, such as anger, happiness, disgust, surprise, sadness, and fear.

The neuron is the basic working unit of the brain, a specialized cell designed to transmit information to other nerve cells, muscles, or gland cells. The human brain has around one hundred billion neurons. There are more potential connections between the neurons than there are atoms in the universe. A fetus grows neurons at the rate of 250,000 a minute. A baby is born with all the neurons of an adult. However, the neural networks are not matured yet. Information travels at different speeds within types of neurons, with transmission rates ranging from three to three hundred feet per second. Figure 15 shows the picture of a neuron. A typical neuron consists of a cell body (soma), dendrites, and axon.

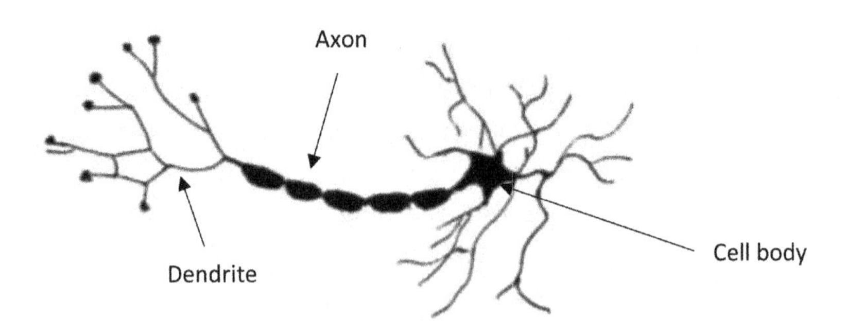

Figure 15: Neuron

Dendrites are thin structures that arise from the cell body, often extending for hundreds of microns and branching multiple times, giving rise to a complex "dendritic tree." An axon is a special cellular extension (process) that arises from the cell body at a site called the axon hillock.

Synapse is a communication site when neurons pass nerve impulses among themselves between the axon of one neuron and a dendrite or soma of another. Axons can run up to several feet before branching. Each neuron has about ten thousand synapses, and there are around one hundred billion neurons. Hence, the number of connections could be around one hundred trillion. This gives an idea of the brain's computing power.

Neurons communicate by chemical and electrical signals in a process known as neurotransmission, also called synaptic transmission. The fundamental process that triggers the release of neurotransmitters is the action potential, a propagating electrical signal that is generated by chemical pulses called ions, which have electrical charges and are made of sodium (Na^+), potassium (K^+), chloride (Cl^-), and calcium (Ca^{2+}). Most impulses are of 100mV and 1mSec duration but are travelling at varying speed.

Brain size or weight does not correlate with intelligence. The average human brain weight is about 1,500 grams. Larger heads may have larger brains. But again, that does not necessarily correlate with intelligence. Einstein's brain weighed in at about 1,230 grams, which is about 18 percent below average. He did have a somewhat wider than normal parietal lobe, which may have contributed to his mathematical genius.

Neural Correlates of Consciousness

Neural correlates of consciousness (NCC) is offered by Crick and Koch. The basic idea is that mental states become conscious when large numbers of neurons all fire in synchrony with one another (oscillations within the thirty-five to seventy-five hertz range or thirty-five to seventy-five cycles per second). Currently, one method used is simply to study some aspect of neural functioning with sophisticated detecting equipment (such as MRIs and PET scans) and then correlate it with first-person reports of conscious experience. Magnetic resonance imaging (MRI) is a medical imaging technique used in radiology to form pictures of the anatomy and the physiological processes of the body in both normal health and disease. MRI scanners use strong magnetic fields, radio waves, and field gradients to generate images of the organs in the body. A positron emission tomography (PET) scan is an imaging test using a special dye that has radioactive tracers. These tracers are injected into a vein of the arm. When highlighted under a PET scanner, the tracers help your doctor see how well your organs and tissues are working.

Another method is to study the difference in brain activity between those under anesthesia and those not under any such influence. A detailed survey would be impossible to give here, but a number of other candidates for the NCC have emerged over the past two decades. These include reentrant cortical feedback loops in the neural circuitry throughout the brain by Edelman and Tononi, NMDA (the N-methyl-D-aspartate receptor, a glutamate receptor, and an ion channel protein found in nerve cells), mediated transient neural assemblies by Flohr, and emotive somatosensory hemostatic processes in the frontal lobe by Damasio. To elaborate briefly on Flohr's theory, the idea is that anesthetics destroy conscious mental activity because they interfere with the functioning of NMDA synapses between neurons, which are those that are dependent on N-methyl-D-aspartate receptors. Ongoing scientific investigation is significant and an important aspect of current scientific research in the field.

Anesthetics do not suppress all brain activities, just the activities associated with consciousness. Unconscious processing like breathing, heart rate, and digestive activities are still going on, and these are controlled by the medulla in the brain stem and spinal cord. Although we normally combine creativity with intelligence, it is not necessarily true. Some very creative people score poorly on intelligence, and some very intelligent people appear to be unable to think "outside the box." The ultimate manifestation of intelligence is in consciousness. Consciousness is the explicit awareness of our experiences and thoughts. Consciousness tends to be associated with language, allowing us to label and categorize the happenings in the world. Although we have impulses and desires to do something, our consciousness is always paramount to determine our actions. Consciousness does not reside in any specific area of the brain. Neither brain has consciousness neurons. Several areas of the brain are responsible for consciousness. These parts include the thalamus, prefrontal cortex, and portions of parietal and temporal lobes.

The phenomenon of consciousness is its unity. Each of us feels that we are a single, unique conscious being who always knows what he knows and doesn't know what he doesn't. Although impulses and desires may pop into our minds, consciousness has the final choice or veto power over our actions. Awareness has a level of perception that can be felt as an experience. However, it lacks, in relation to consciousness, mostly language, by which the experience is interpreted in a rational scheme. Consciousness relates and catalogues the experience in rational scheme in memory. Nobel Prize winner Gerald Edelman has referred to consciousness as the "remembered present." This remembering is embodied in language.

Brain Waves

At the root of all our thoughts, emotions, and behaviors is the communication between neurons within our brains. Brain waves are produced by synchronized electrical pulses from masses of neurons communicating with each other. Brain waves are rhythmic or repetitive neural activity in the brain. Neural tissue can generate oscillatory activity in many ways, driven either by mechanisms within individual neurons or by interactions between neurons. The interaction between neurons can give rise to oscillations at a different frequency than the firing frequency of individual neurons. In general, oscillations can be characterized by their frequency, amplitude, and phase. Along with one hundred billion neurons and one hundred trillion constantly changing connections, individual neurons use precise rhythms and groups of neurons oscillating together in specific frequencies. The perplexing relationship of neuronal networks and brain waves is critical to future understanding of the brain.

Neurons generate action potentials resulting from changes in the electric membrane potential. Neurons can generate multiple action potentials in sequence, forming so-called spike trains (figure 16). These spike trains are the basis for neural coding and information transfer in the brain. Spike trains can form all kinds of patterns, such as rhythmic spiking and bursting, and often display oscillatory activity. Oscillatory activity in single neurons can also be observed in subthreshold fluctuations in membrane potential. These rhythmic changes in membrane potential do not reach the critical threshold and therefore do not result in an action potential. They can result from postsynaptic potentials from synchronous inputs or from intrinsic properties of neurons.

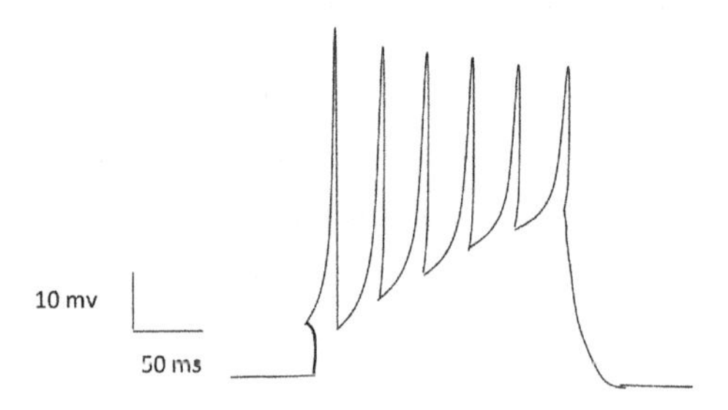

Figure 16: Spike train

Spikes are produced in response to stimuli or spontaneously, and each spike typically lasts for one millisecond. A spike train is simply a combinatorial sequence of spikes and silences. A popular way to think of spike trains is as a digital sequence of information: 1 for a spike, and 0 for no spike. For example, an encoded spike train structure could look like 001111101101. The first two 0s represent the latency time between the stimulus presentation and the first spike. Spike trains can be induced by physical sensory stimuli such as vision, touch, smell, or sound; or they can be generated by abstract stimuli such as cognitive stimulation by evoking memory. The duration and structure of a spike train generally depends upon the intensity and duration of the stimulus. Spike trains often last as long as the stimulus is present. However, some neurons have special electrical properties wherein they generate a sustained response to a short stimulus. In the case of these neurons, greater stimulus intensities are more likely to generate longer spikes.

The major brain wave frequency ranges are as follows:

Infra-low waves (<0.5 Hz): Very little is known about them. They appear to play a major role in brain timing and network function.

Delta waves (0.5–4 Hz): They are generated in deepest meditation and dreamless sleep. They suspend external awareness and are the source of empathy. Healing and regeneration are stimulated in this state. They are found most often in infants as well as young children. As we age, we tend to produce fewer delta wave even during deep sleep. They have also been found to be involved in unconscious bodily functions such as regulating heartbeat and digestion. Adequate production of delta waves helps us feel completely rejuvenated after we wake up from a good night's sleep. If there is abnormal delta activity, an individual may experience learning disabilities or have difficulties maintaining conscious awareness (such as in cases of brain injuries).

Theta waves (4–8 Hz): They occur most often in sleep but are also dominant in deep meditation. They act as our gateway to learning and memory. In theta, our senses are withdrawn from the external world and focused on signals originating from within. Theta waves are connected to our experiencing and feeling deep and raw emotions. Too much theta activity may make people prone to bouts of depression and becoming highly suggestible because they are in a deeply relaxed, semihypnotic state. Theta has its benefits of helping improve our intuition and creativity and making us feel more natural. They are also involved in restorative sleep. As long as theta isn't produced in excess during our waking hours, it is a very helpful brain wave range.

Alpha waves (8–12 Hz): They are dominant during quietly flowing thoughts and in some meditative states. They aid overall mental coordination, calmness, alertness, mind/body integration, and learning. This frequency range bridges the gap between our conscious thinking and subconscious minds. In other words, alpha is the frequency range between beta and theta. They help us calm down when necessary and promotes feelings of deep relaxation. If we become stressed, a phenomenon called "alpha blocking" may occur, which involves excessive beta activity and very little alpha. Essentially the beta waves "block" out the production of alpha because we become too aroused.

Beta waves (12–40 Hz): They dominate our normal waking state of consciousness when attention is directed toward cognitive tasks and the outside world. Beta is a "fast" activity, present when we are alert, attentive, and engaged in problem-solving, judgment, decision-making and focused mental activity. They are further divided into three bands. Beta1 (13–15 Hz) can be thought of as fast idle, or musing. Beta2 (16–22 Hz) is high engagement or actively figuring something out. Beta3 (23–40 Hz) is highly complex thought, integrating new experiences, high anxiety, or excitement. They are involved in conscious thought and logical thinking, and they tend to have a stimulating affect. Having the right amount of beta waves allows us to focus and complete school or work-based tasks easily. Having too much beta may lead to us experiencing excessive stress and/or anxiety. The higher beta frequencies are associated with high levels of arousal. When you drink caffeine or have another stimulant, your beta activity will naturally increase. These are fast brain waves that most people exhibit throughout the day in order to complete conscious tasks such as critical thinking, writing, reading, and socialization.

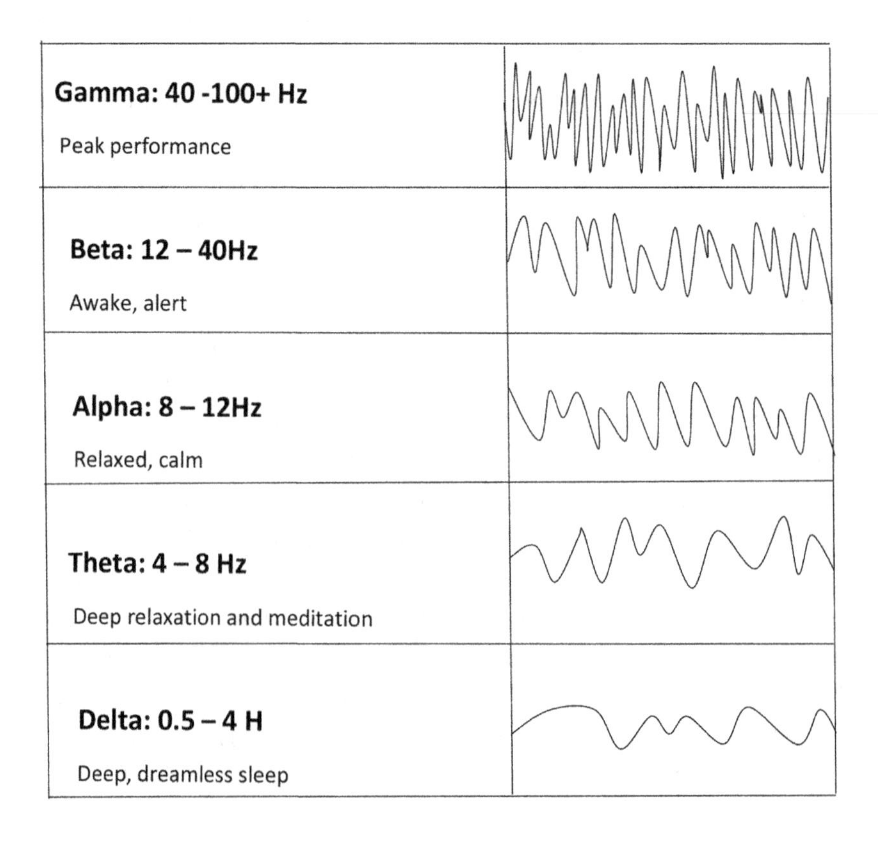

Figure 17: Brain waves at different frequency range

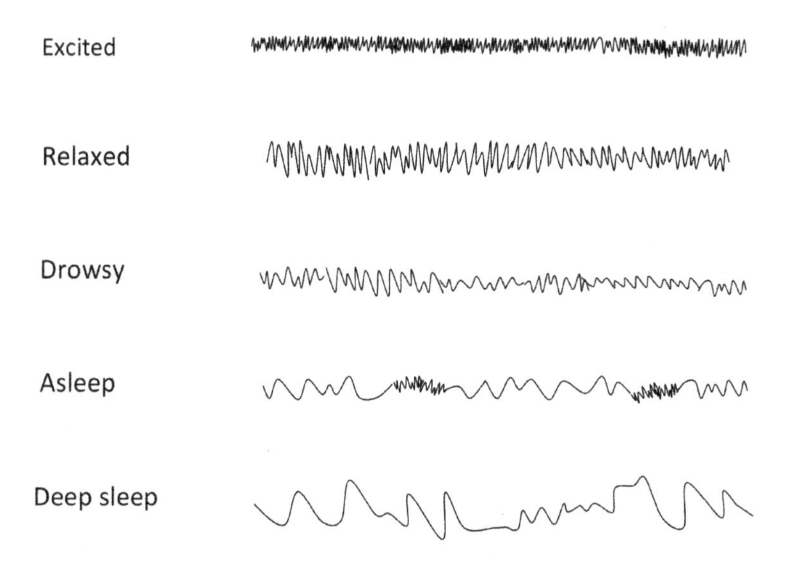

Figure 18: Brain waves in different conditions

Gamma waves (40–100 Hz): They relate to simultaneous processing of information from different brain areas. They pass information rapidly and are the subtlest of the brain wave frequencies. The mind has to be quiet to access it. These are involved in higher processing tasks as well as cognitive functioning. Gamma waves are important for learning, memory, and information processing. It is thought that the 40 Hz gamma waves are important for the binding of our senses regarding perception and are involved in learning new material. It has been found that individuals who are mentally challenged and have learning disabilities tend to have lower gamma activity than average. The brain waves are shown in figure 17 and brain waves in different conditions in figure 18.

Gamma brain waves are considered the brain's optimal frequency of functioning. Gamma brain waves are commonly associated with increased levels of compassion, feelings of happiness, and optimal brain functioning. Gamma brain waves are associated with a conscious awareness of reality and increased mental abilities. A gamma wave is a pattern of brain waves associated with perception and consciousness.

The dominant power spectral distribution in healthy subjects is in the range frequency of delta and theta. The 10–20 system is an internationally recognized method to describe and apply the location of scalp electrodes in the context of an EEG test or experiment. The highest value of power spectral is 263.3 μV^2 and is in the frequency 3 Hz. This indicates that the right anterior frontal lobe experiences the highest alpha wave activity when recording the resting state. The power spectral for schizophrenic patients has varying value. There are significant differences between schizophrenic and normal subjects in the delta waves in the frontal lobe, theta waves in frontal lobe, and in the alpha waves in the occipital lobe. There are no significant differences between schizophrenic and normal in the beta and gamma waves.

The spectral power for normal subjects has the highest value in the frequency range 10.25–11 Hz. The highest power spectral power of schizophrenic patients is in the frequency range of 3–3.5 Hz. We do not tend to experience solely one kind of brain wave at any given time. In other words, it is inaccurate to say a person is experiencing "alpha waves," for in fact they will be displaying all five types of brain waves at once. Rather, it is more accurate to say that one type of brain wave is dominant. What does this mean? It means that some parts of the brain will be buzzing more than others are at any given time, but we can take an average in order to estimate roughly how someone is feeling. The dominant state indicates "state of mind" or level of consciousness. Because different areas of the brain may have different activity at any given time, there will be one activity in one brain wave state in one area of the brain, while at the same time, a different brain wave state may be more active in another area of the brain. Each of these brain wave states occurs in a specific frequency range.

Measurement of Brain Waves

Doctors and researchers use brain-imaging techniques to view activity or problems within the human brain without invasive neurosurgery. There are a number of accepted safe imaging techniques in use today in research facilities and hospitals throughout the world.

fMRI

Functional magnetic resonance imaging, or fMRI, is a technique for measuring brain activity. The fMRI concept builds on the earlier MRI scanning technology and the discovery of properties of oxygen-rich blood. The cylindrical tube of an MRI scanner houses a powerful electromagnet. A typical scanner has a strength of three teslas, about fifty thousand times greater than the earth's magnetic field. The magnetic field of the scanner affects the magnetic nuclei of atoms. Normally atom nuclei are randomly oriented, but under the influence of a magnetic field, the nuclei become aligned with the direction of the magnetic field. The stronger the field, the greater the degree of alignment. When pointing in the same direction, the tiny magnetic signals from the individual nuclei add up coherently, resulting in a signal that is large enough to measure. In fMRI, the magnetic signal from hydrogen nuclei in water (H_2O) is detected. The key to fMRI is that the signals from hydrogen atoms vary in strength depending on the surroundings. This provides a means for discriminating between gray matter, white matter, and cerebral spinal fluid in structural images of the brain. This works by detecting the changes in blood oxygenation and flow that occur in response to neural activity—when a brain area is more active, it consumes more oxygen, and to meet this increased demand, blood flow increases to the active area. The fMRI can be used to produce activation maps showing which parts of the brain are involved in a particular mental process.

What is the difference between MRI and fMRI?

MRI is designed for revealing the peculiarities of anatomical structure inside a human organism, including those in the human brain. On the other hand, fMRI maps the image via measuring the blood flow levels in the human brain. The data captured with the fMRI shows changes in the metabolic functioning in the brain.

Due to the different ways of measuring processes in the human brain, MRI and fMRI differ in terms of the resulting picture. MRI measures the molecule called hydrogen nuclei. The captured data allows MRI to create a spatial image of the finest resolution of the human brain. The fMRI measures oxygen levels flowing into the brain and calculates the differences in tissue with respect to time.

MRI technique has established a solid reputation in modern medicine. It is widely used in various fields of medicine as well as in medical studies. The fMRI, on the other hand, is a relatively recent technique that is only beginning to gain popularity.

CT

Computed tomography (CT) scanning builds up a picture of the brain based on the differential absorption of X-rays. During a CT scan, a patient lies on a table that slides in and out of a hollow cylindrical apparatus. A narrow beam of X-rays is aimed at the patient and quickly rotated around the head, producing signals that are processed by the machine's computer to generate cross-sectional images—or "slices"—of the head. Images made using X-rays depend on the absorption of the beam by the tissue it passes through. Bone and hard tissue absorb X-rays well, air and water absorb very little, and soft tissue is somewhere in between. Thus CT scans reveal the gross features of the brain but do not resolve its structure well.

PET

A brain positron emission tomography (PET) scan is an imaging test of the brain. It uses a radioactive substance called a tracer to look for disease or injury in the brain. A PET scan shows how the brain and its tissues are working. Other imaging tests, such as magnetic resonance imaging (MRI) and computed tomography (CT) scans, only reveal the structure of the brain.

A PET scan requires a small amount of radioactive material (tracer). This tracer is given through a vein, usually on the inside of elbow or breathing in the radioactive material as a gas. The tracer travels through blood and collects in organs and tissues. The tracer helps the doctor see certain areas or diseases more clearly.

The patient waits nearby as his body absorbs the tracer. This usually takes about an hour. Then the patient lies on a narrow table that slides into a large tunnel-shaped scanner. The PET scanner detects signals from the tracer. A computer changes the results into 3-D pictures. The images are displayed on a monitor for the doctor to read.

Positron emission tomography (PET) uses trace amounts of short-lived radioactive material to map functional processes in the brain. When the material undergoes radioactive decay, a positron is emitted, which can be picked up by the detector. Areas of high radioactivity are associated with brain activity.

EEG

Electroencephalography (EEG) is the measurement of the electrical activity of the brain by recording from electrodes placed on the scalp. It is a noninvasive test, with the electrodes placed along the scalp, although invasive electrodes are sometimes used in specific applications. The EEG measures voltage fluctuations resulting from ionic current within the neurons of the brain. Over a period of time, EEG records the brain's spontaneous electrical activity from multiple electrodes placed on the scalp. Diagnostic applications generally focus on the spectral content of EEG—that is, the brain waves observed in the EEG signals.

The resulting traces are known as an electroencephalogram (EEG) and represent an electrical signal from a large number of neurons. EEGs are frequently used in experimentation because the process is noninvasive to the research subject. The EEG is capable of detecting changes in electrical activity in the brain on a millisecond level. It is one of the few techniques available with such high temporal resolution. Temporal resolution refers to the precision of a measurement with respect to time.

MEG

Magnetoencephalography, or MEG scan, is an imaging technique that identifies brain activity and measures small magnetic fields produced in the brain. The scan is used to produce a magnetic source image (MSI) to pinpoint the source of seizures. Magnetic fields are detected by arrays of SQUIDs (superconducting quantum interference devices). It doesn't emit radiation or magnetic fields.

These measurements are commonly used in both research and clinical settings. There are many uses for the MEG, including assisting surgeons in localizing pathology, assisting researchers in determining the function of various parts of the brain, neurofeedback, and others. Results from the MEG test are matched up with a magnetic resonance image (MRI), which is an anatomical picture of the brain. The MEG and MRI create a "map," or magnetic source image (MSI), that shows areas of normal and abnormal activity in the brain.

fNIRS

Functional near-infrared spectroscopy (fNIRS) is a noninvasive optical imaging technique that uses low levels of light to measure blood flow changes in the brain associated with brain activity, such as performance of a task. It works by shining light in the near infrared part of the spectrum (700–900 nm) through the skull and detecting how much the remerging light is attenuated. How much the light is attenuated depends on blood oxygenation, and thus fNIRS can provide an indirect measure of brain activity. It has helped to advance basic science brain-mapping studies, identifying areas of the brain associated with a range of motor and visual tasks. The fNIRS systems provide noninvasive measurements of oxygen saturation and hemoglobin.

Experiments on Near-Death Brain Activities

Experiments by Borjigin, Lee, Liu, Pal, Huff, Klarr, Sloboda, Hernandez, Wang, and Mashour on rats have found a surge of gamma waves—that is, brain waves of frequencies in gamma range greater than 40 Hz within the first thirty seconds of cardiac arrest. This was followed by a second stage of six seconds of delta waves—that is, brain waves of frequencies less than 4 Hz. The third stage has brain waves of gamma range. The final stage has brain waves of frequencies in ultra-gamma range (300 Hz). This shows that after clinical death, the brain goes through a period of heightened brain activity found in normal consciousness. In another experiment, Loretta Norton and colleagues examined EEG recordings of four critically ill patients at the point where their life support was withdrawn. They found no evidence of the large delta or gamma blips that has been observed in rats. However, it needs many further experiments.

Brain Wave Growth in Children

As mentioned, the average human brain weighs about 1,500 grams. Einstein's brain was about 10 percent less than average, about 1,230 grams. However, he had a wider than normal parietal lobe, which might have contributed to his mathematical ability. Brain growth starts in the very early stage of pregnancy, with a sharp increase from twenty to thirty weeks of pregnancy. A baby is born with one hundred billion neurons, the same number an adult brain has. It has also some synapses (connections between the neurons). An adult brain has a quadrillion synapse—that is, one followed by fifteen zeros (1,015). At birth, the number of synapses per neuron is about 2,500. Therefore, the total number of synapses is 2.5 x 1012. But by age two or three, it's about the same as an adult's.

The brain waves in a baby from birth to two years are in the lowest brain frequency range, 0.5 to 4 cycles per second, and are called delta waves. This is the same range of an adult in deep sleep or meditation. For children between two to six years, the brain wave frequency is in the theta range of four to eight cycles per second. This corresponds to an adult in a deep, relaxed state. For children between six and eight, the brain waves move into an alpha range of frequency of eight to twelve cycles per second. This corresponds to an adult in mental coordination and alertness. From eight years onward, the brain waves move into beta range from twelve to forty cycles per second for conscious and analytical thinking as an adult brain.

Hence, babies and children should be positively helped and encouraged to grow their brains. Their inquisitiveness and questions should be answered and supported with positive answers and actions. In that respect, parents, especially mothers, have an important responsibility for the growth of the children. Tests have shown that at birth a baby is already familiar with the mother's voice from hearing it in the womb. This is because the neurons in the auditory cortex of the baby are responding to the speech sounds of the mother.

Brain Change with Age

Many studies have found that intelligence and cognitive ability do not inevitably decline with age. Intellectual practice is crucial for maintaining and even developing cognitive skills during aging. Problem-solving is an important exercise for this. Good nutrition contributes to the maintenance of intellectual abilities. Physical exercise increases brain rhythms associated with memory and concentration. It also stimulates the growth of neurons and removal of damaged neurons. There appears to be no intelligence pills that are safe and effective. Vitamin supplements may improve the functions of the body and improve cognitive abilities along with health and alertness.

Brain Waves in Alzheimer's and Schizophrenic People

Alzheimer's disease is commonly observed in the aged. But it is not solely a disease of the elderly, nor is it inevitable as you age. There are very young Alzheimer's victims and some very old people without Alzheimer's. Gamma brain waves—electrical charges that help link and process information from all parts of the brain—are known to slow down in the brains of people with Alzheimer's disease and other neurological or psychiatric disorders. Neurons that die in the early disease course are particularly in the hippocampus because this structure is vital for transferring memories from short term to long term. Scientists have seen disrupted gamma waves in many types of brain disorders, including injuries, schizophrenia, and Alzheimer's disease. By restoring normal gamma waves, scientists actually managed to counteract a hallmark of the disease. In Alzheimer's disease, a protein called beta-amyloid gathers in the spaces between neurons and creates large, harmful plaques. However, gamma rays can apparently mobilize the immune system to clear these plaques.

What Are Physical Constants?

A physical constant, sometimes called fundamental physical constant, is a physical quantity that is generally believed to be both universal in nature and have constant value in time. There are many physical constants in science; some of the most widely recognized being the gravitational constant G, speed of light in vacuum c, Planck's constant h, Boltzmann constant k_B, the vacuum permittivity constant ε_0, permeability constant μ_0, the elementary charge e, and so forth. The question is why we need a physical constant. Let us take gravitational force or gravity. Every mass produces gravitational gravity force. The sun has gravitational force; Earth has gravitational force. But the sun's mass is about 333,000 times more than the earth's mass, hence the sun's gravitational force. The earth goes around the sun. Earth's mass is about 6 x 1,022 more than the average human mass. That's why if we jump up, we immediately come down to earth. Physics cannot explain why mass has gravity. However, the gravitational force can be calculated by using a physical constant called gravitational constant G, which has a certain value. The gravitational force F between two masses m_1 and m_2 is given by $F = G (m1\ m2)/r^2$, where G = gravitational constant = $6.67259 \times 10^{-11}\ m^3\ Kg^{-1}\ s^{-2}$, m_1 and m_2 are the two masses, and r is the distance between them.

Similarly, light consists of photons. Light has energy; hence, photon has energy. But physics cannot explain where photon gets its energy from. Blue light has more energy than red light; ultraviolet ray has more energy than infrared ray. For that reason, photons in blue light have more energy than photons in red light; photons in an ultraviolet ray have more energy than an infrared ray. Planck solved this problem by using Planck constant h. Planck constant = h = 6.626 $\times 10^{-34}$ m^2 kg / s.

According to his formula, energy E of a photon is given by E = h.f, where h = Planck constant and f is frequency. The part of spectrum of light from low to high frequency is infrared, red, orange, yellow, blue, indigo, violet, ultraviolet. Hence, the energy of infrared photon is lower than red, blue, and so on.

Physics cannot explain why gas has energy. But by using a physical constant called Boltzmann constant (k_B), the average kinetic energy (E) of a particle in a gas at an absolute temperature (T) can be determined: $E = k_B.T$; $k_B = 1.380658 \times 10^{-23}$ joules per kelvin ($J \cdot K^{-1)}$.

The force F between two separated electric charges in the vacuum of classical electromagnetism is given by Coulomb's law: $F = 1/(4\pi\ \varepsilon_0)$ $(q_1\ q_2)/r^2$. Where ε_0 is a physical constant called permittivity constant, $\varepsilon_0 = 8.854 \times 10^{-12}$ farad per meter, q_1 and q_2 are the electric charges and r is the distance between them.

A list of physical constants is given in appendix 1.

Theory of Consciousness as Physical Constant

I have proposed in my article "Consciousness As a Function of Brain Waves and Physical Constant *Conscire*," published in *NeuroQuantology* (September 2017, volume 15, issue 3), that consciousness is derived from a fundamental physical constant *Conscire C* working on brain waves. *Conscire* is a Latin word for conscious derived from *com*, "with" or "thoroughly," *sciere*, to "know."

Rhythm	Freq Hz	Mental Condition	Consciousness
Delta	0.5– 4	Dreamless sleep and meditation	Very low
Theta	4–8	Deeply relaxed and semihypnotic state	Low
Alpha	8–12	Mental coordination, calmness, and alertness	Medium
Beta	12–40	Attentive, engaged in problem-solving, judgment, decision-making, and focused mental activity.	High
Gamma	40–100	Cognitive functioning, important for learning, memory, and processing	Very high

Table 1

As described before, brain waves are given in Table 1. These waves are identified by frequency (Hz or cycles/sec) and amplitude. The amplitudes recorded by scalp electrodes are in the range of microvolts (μV or 1/1,000,000 of a volt). Table 1 also shows the state of consciousness in different rhythm or brain wave frequency range.

As seen in Table 1, frequency of the brain wave changes with the mental activity. Hence, mental activity is a function of brain wave frequency. When the brain wave frequency is high, the brain is very active. When we are in a state of deep sleep or meditation, the brain activity is very low and brain wave frequency is 4 Hz and less. When we are deeply relaxed, then the brain activity is low and the brain wave frequency is between 4 Hz to 8 Hz. When we are calm and alert, our brain activity is getting higher and the brain wave frequency is between 8 and 12 Hz. When we are engaged in mental activities like problem-solving, decision-making, and so forth, our brain activity is high and brain wave frequency is between 12 Hz and 40 Hz. When we are engaged in extremely busy mental function, our brain activity is very high and the brain frequency is between 40 Hz and 100 Hz. It is therefore obvious that brain activity is proportional to the brain wave frequency (BWF).

Brain activity needs brain energy. Hence, brain energy is proportional to the brain wave frequency. Consciousness is related to brain activity, hence brain energy. Consciousness is therefore proportional to the brain wave frequency. If physical constant of consciousness is Consciere C, then consciousness is equal to C.f.

As gravitational constant G is involved in the gravitational force between two masses, Boltzmann constant k_B is involved in the heat of a gas particle, Planck constant h is involved in determining the energy of a light wave, and similarly Conscire C is involved in determining consciousness created by brain waves. Just as gravitational constant G needs matter, Boltzmann constant needs temperature, Planck's constant h needs a wave frequency, electric constant ε_0 needs electric charge, and similarly Conscire C needs brain waves and works only on brain waves. Consciousness per brain wave frequency f for unit amplitude = C.f

Hence, consciousness needs a brain. If the brain is dead, then there is no consciousness. Now, what about other living mammals, vertebrates like lizards, birds, and so on? Do they have consciousness? Only mammals have a cortex in the brain. Other vertebrates have only basal ganglia, which is a group of subcortical nuclei in the brain. Only humans have the most developed cortex, with half of the cortex as frontal.

Analysis of Brain Waves

Brain waves consist of dominant waves in one of the ranges (delta, theta, alpha, beta, gamma) depending on mental activity and nondominant waves in other ranges, as shown in figure 18. This waveform is in time domain. To analyze it, this should be converted into frequency domain. The Fourier transform is the mathematical tool used to make this conversion. Simply stated, the Fourier transform converts waveform data in the time domain into the frequency domain.

As a musical chord can be expressed as the frequencies (or pitches) of its constituent notes, similarly the Fourier transform converts a time signal into the frequencies that make it up. The Fourier transform is called the frequency domain representation of the original signal. A waveform in time domain is difficult to describe mathematically. However, its representation in sinusoidal waveforms of different frequencies is easy to describe and manipulate. Fourier transform comes from the study of Fourier series, which represents a function as the sum of simple sine waves. Fourier transform decomposes any periodic function into the sum of a set of simple oscillating functions, namely sines and cosines. Figure 19 shows the waveforms of different waveforms in time domain and their frequency spectrum in frequency domain.

Numerical calculation of a Fourier transform is a tremendously labor-intensive task because such a large amount of arithmetic had to be performed with paper and pencil. Software algorithms have been developed for personal computers for Fourier transform analysis. An alternative to the FFT is the discrete Fourier transform (DFT). The DFT allows a precisely defined range over which the transform will be calculated, which eliminates the need to window. On the negative side, the DFT is computationally slower than the FFT. The original signal can be reconstructed as a function of time by computing the inverse Fourier transform (IFT) from the power spectrum displayed by one of the two previously mentioned transforms.

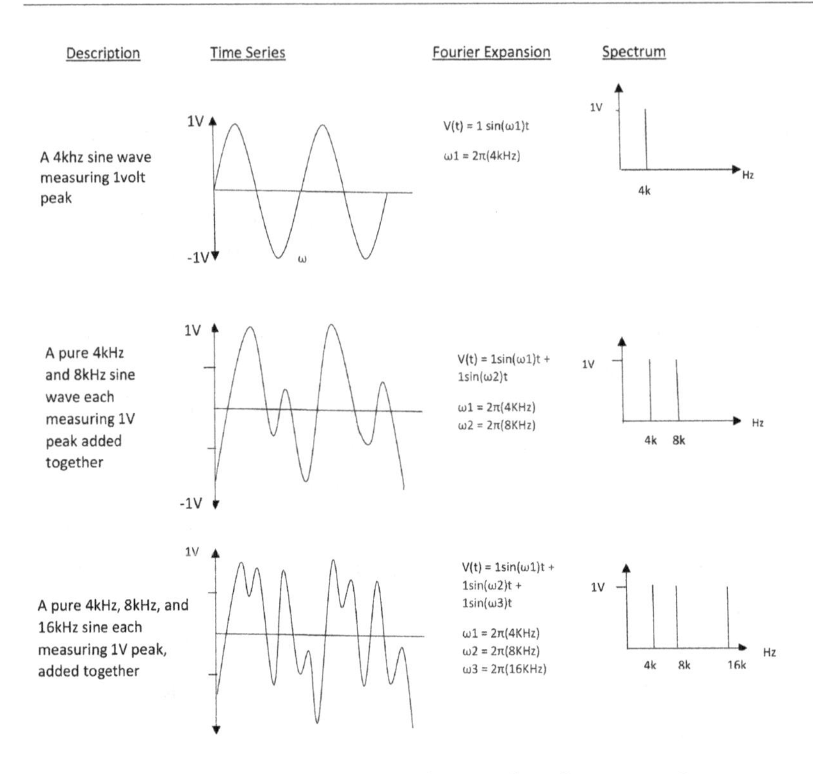

Figure 19: Fourier transforms of various waveforms

The FFT is just a faster implementation of the DFT. FFT on 1,024 (i.e., 210) data points would require (n/2) $\log_2$ (n) = 512 × 10 = 5,120 multiplications, where n = 210. DFT on 1,024 (i.e., 210) data points would require n^2 = 1,024 × 1,024 = 2^{20} = 1,048,576, which is a factor of 205 complexity.

Because of the base 2 logarithm, by definition FFT function contains a total number of data points precisely equal to a 2-to-the-nth-power number (e.g., 512, 1,024, 2,048, etc.). Therefore, with an FFT, only a fixed length of waveform containing 512 points, or 1,024 points, or 2,048 points, and so forth, can be evaluated. If the time series contains 1,096 data points, only 1,024 of them at a time using an FFT can be evaluated since 1,024 is the highest 2-to-the-nth-power that is less than 1096.

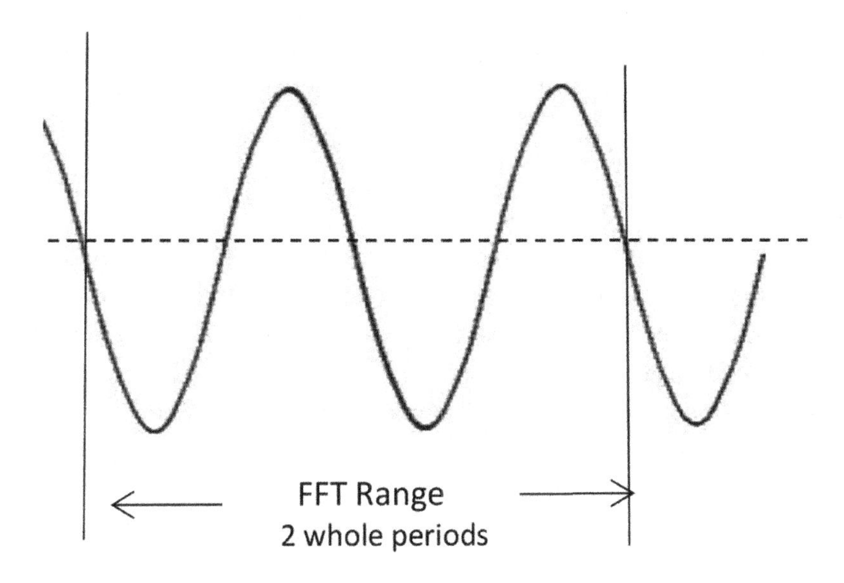

Figure 20(a): Whole number of periods

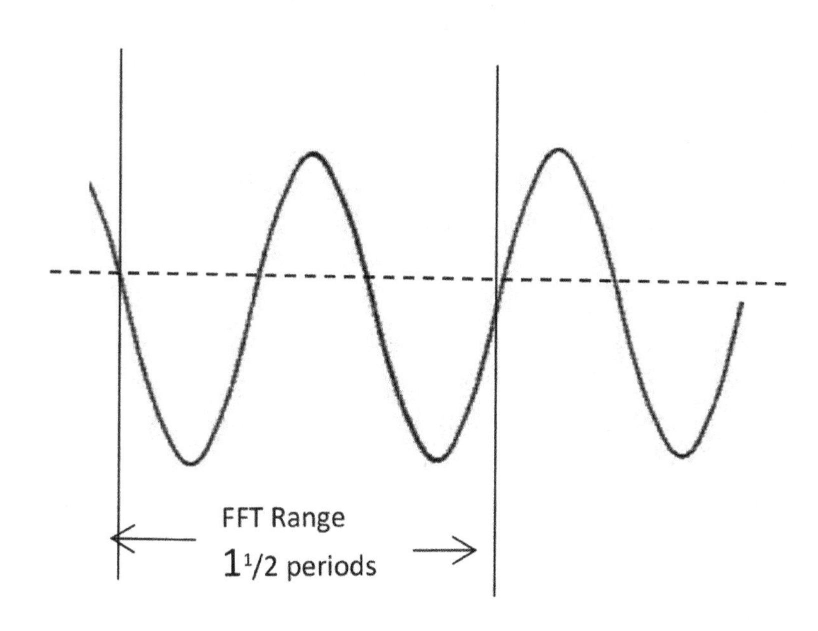

Figure 20(b): Fractional number of periods

Figure 20(a) shows a best-case waveform, which is not common, where the range of the FFT exactly contains a whole number of periods, starting with the waveforms mean value. This waveform possesses end point continuity, which means the resulting power spectrum will be accurate and no window need be applied. A more typical case is shown in figure 20(b), where the range of the FFT does not contain a whole number of periods. The discontinuity in the endpoints of this waveform means the resulting power spectrum will contain high-frequency components not present in the input, requiring the application of a window to attenuate the discontinuity and improve accuracy. Thus the FFT would evaluate this waveform with the endpoint error and generate a power spectrum containing false frequency components representative of the endpoint mismatch.

While processing transient signals, the edges may contain important information that will be unacceptably distorted by applying the window solution. In this case, there is no choice but to use the DFT. As stated previously, the DFT allows adjusting of the end points that define the range of the waveform to be transformed, thus eliminating the need for windowing. This approach allows a waveform containing any number of points to be evaluated, which provides more flexibility than the fixed-length 2-to-the-nth-power FFT. However, to prevent the same leakage effect experienced with a nonwindowed FFT, the DFT must be generated over a whole number of periods starting at the waveforms mean-level crossing. In other words, the endpoints that define the range of the waveform over which the DFT will be calculated must be adjusted to enclose or define a whole number of periods, preferably starting at or around the point where the waveform crosses its mean.

The DFT allows more versatility and precision than the FFT. However, versatility and precision come at the expense of added computation time by the algorithm and added time spent on endpoint positioning. Inverse Fourier transform (IFT) reconstructs the original signal as a function of time from the power spectrum as a function of frequency by backward run.

Calculation of Brain Wave Energy

Let us analyze brain waves of a person in peak performance. In this case, the dominant brain waves will be in gamma range, with additional waves in beta range and some in alpha range. Let us call the brain wave frequencies in gamma range f_{g1}, f_{g2}, f_{g3}, f_{g4}, and f_{g5} and their amplitudes A_{g1}, A_{g2}, A_{g3}, A_{g4}, and A_{g5}. Similarly, brain waves in beta range as f_{b1}, f_{b2}, and f_{b3} and their amplitudes A_{b1}, A_{b2}, and A_{b3}, and brain wave frequency in alpha range f_{a1} and its amplitude A_{a1}.

As stated earlier, consciousness energy unit per brain wave frequency f for unit amplitude = *C*.f, where the conscious constant is *Consciere*. Hence the total brain wave energy BWE would be as follows:

$$\begin{aligned} BWE &= C.A_{g1}^{2}.f_{g1} + C.A_{g2}^{2}.f_{g2} + C.A_{g3}^{2}.f_{g3} + C.A_{g4}^{2}.f_{g4} + C.A_{g5}^{2}.f_{g5} \\ &\quad + C.A_{b1}^{2}.f_{b1} + C.A_{b2}^{2}.f_{b2} + C.A_{b3}^{2}.f_{b3} + C.A_{a1}^{2}.f_{a1}. \\ &= C\,(A_{g1}^{2}.f_{g1} + A_{g2}^{2}.f_{g2} + A_{g3}^{2}.f_{g3} + A_{g4}^{2}.f_{g4} + A_{g5}^{2}.f_{g5} + A_{b1}^{2}. \\ &\quad f_{b1} + A_{b2}^{2}.f_{b2} + A_{b3}^{2}.f_{b3} + A_{a1}^{2}.f_{a1}) \end{aligned}$$

.... (Eqn. 8.1)

, considering that energy of a wave is proportional to the square of its amplitude.

If we can measure the brain energy and can plot its wave in frequency domain, then we can find out the value of the consciousness constant *Conscire C*. The brain wave energy of brain waves is so weak that it is hardly measurable at all. For comparison, says Dimitrios Pantazis, director of the Magnetoencephalography (MEG) Laboratory at MIT's McGovern Institute, "The magnetic field of the earth is just strong enough to move the needle of a compass. Signals from the brain are a billionth of that strength." As technology advances, it would be possible to measure such weak signal. Then the value of consciousness constant *C* in joules (X) or in electron volt (eV) (Y) can be determined as follows:

C = X joules or Y eV per Hz. 1 joule = 1 Kg m^2 s^{-2} = 6.24 x 1018 eV, hence C = X Kg m^2 s^{-2} or Y eV per Hz.

The question that might arise is that these frequencies might come from other sources. Does it mean that these sources will have consciousness? No, the answer is that the source must be the brain. It is true for any other application in science. Let us consider cosmic microwave background (CMB). These are radiations in the universe in the frequency range from 0.3 GHz to 630 GHz at a peak frequency of 160.4 GHz. This has confirmed the big bang theory. Does it mean that any microwave radiation in that frequency range from any source will confirm big bang theory? No, the source must be the universe. Similarly, when we are listening to a maestro playing a piano, the acoustic wave frequency must be from the piano, not from any other source. This applies to every application, as is true for brain waves for consciousness.

Conclusion

Consciousness has been a mystery since early civilization. So far, religious and philosophical explanations could not solve the mystery. There have been a dualistic view and monistic view of consciousness. Dualism believes that mind or consciousness is different from body and brain. Monism believes that consciousness is a part of the brain. This is the most dominant view of the current scientists. Quantum mechanics has been particularly intriguing to provide an explanation of consciousness. Quantum theory and consciousness are both poorly understood; what they have in common is degree of mystery and magic. Most neurologists think that the current approach in physics to explain consciousness would not work. It would only make it more complex and confusing. But what is consciousness? What substance is it made of? No current theory of brain can explain where and how consciousness happens. Few philosophers have proposed consciousness as fundamental property. David Chalmers's theory of consciousness proposes consciousness as a fundamental property like space, time, mass, charge, and so forth. Integrated information theory proposes consciousness as fundamental property of a physical system. Russellian monism states that consciousness is the intrinsic nature of brain states.

The best and only way out of this dead end is to accept that consciousness must be physical property of brain waves. The logic is as follows:

> Science has proven that brain waves and brain energies exist.
>
> A person who is brain dead is not conscious, and brain waves and brain energies do not exist.
>
> Brain waves and brain energies exist only for living objects with the brain.
>
> The nature and state of consciousness depends on the frequency of the brain waves.

Hence, it can be concluded that consciousness is a fundamental property of brain waves. Newton introduced gravitational constant as a fundamental property of mass to solve gravity. Planck introduced Planck constant as a fundamental property of photon to solve blackbody radiation. Boltzmann introduced Boltzmann constant to explain the kinetic energy of gas particles with temperature. Coulomb introduced permittivity constant to calculate the force between electric charges. Similarly, consciousness constant *Conscire C* is a constant of brain waves to explain consciousness. If there is no brain wave, there is no consciousness. The value of consciousness constant *Conscire C* could not be determined since current technology could not measure the value of brain energy. But as technology advances, it would be possible to measure the brain energy, hence determine the value of consciousness constant *Conscire C* by calculating the frequencies and amplitudes of the measured brain waves obtained from DFT analysis and applying equation (9.1) against the measured brain energy.

EPILOGUE

Consciousness has been an intriguing and puzzling subject since early civilization. It has been analyzed from subjective and objective aspects, monism and dualism aspects, classical physics and quantum mechanism aspects, religious aspect, and philosophical aspect. However, it remains a mystery. Scientists have realized that objective measurement and verification could no longer be the mark of absolute reality because the measured object could no longer be completely separated from the measuring subject—the measured and the measurer, the verified and the verifier, are the same. As Erwin Schrödinger said, "Subject and object are only one. The barrier between them cannot be said to have broken down as a result of recent experience in the physical science, for this barrier does not exist." Heisenberg concluded, "The common division of the world between subject and object, inner world and outer world, body and soul, is no longer valid and leads to difficulties." Schrödinger concurred and stated simply, "These shortcomings can hardly be avoided except by abandoning dualism."

The essential insight of the work of Heisenberg, Schrödinger, and Einstein is that the texture of reality is one in which the subject and the object, observer and the event, and the knower and the known are not separable. Taoism recognizes these two general forms of knowing as conventional knowledge and natural knowledge—that is, the knowledge of the universe as it is conventionally defined and named as opposed to the knowledge of the way (tao) the universe is in its actuality.

Mundaka Upanishad in Hinduism clearly states these two modes of knowing. The lower mode, termed aparavidya, corresponds to symbolic-map knowledge, and is based on the distinction between the knower (paramatr) and the known (visaya). The higher mode, called paravidya, is reached not through the lower mode, but all at once intuitively. It is a unique and intuitive vision of nonduality.

Christian dualism refers to the belief that God and creation are distinct but interrelated through an indivisible bond. Throughout much of the history of Christian thought, Christians have held to the idea that there is a very strict separation between soul and body, mind and brain.

In Buddhism, vijnana signifies the mode of knowledge, which is dualistic, and prajna is the mode of knowledge which is nonconceptual, nonsymbolic, and nondual. David Suzuki says, "Prajna goes beyond vijnana. We use vijnana in the world of the senses and intellect, which is characterized by dualism in the sense that there is one who sees and there is the other that is seen—the two standing in opposition. In prajna, this differentiation does not take place; what is seen and the one who sees are identical; the seer is the seen and the seen is the seer."

Many great philosophers and thinkers were intrigued by consciousness and gave their own theories of its origin, whether dualistic, monistic, or otherwise. However, there was no consensus or conclusion. As science progressed, we came to know more about brain structure and its function and how synapses of one neuron passes brain signals to another neuron. But consciousness still remained a mystery.

Advent of quantum mechanics changed the scenario. Quantum mechanics (also known as quantum physics or quantum theory), including quantum field theory, is a branch of physics that is the fundamental theory of nature at small scales and low-energy levels of atoms and subatomic particles. Quantum mechanics differs from classical physics in that energy, momentum, and other quantities are often restricted to discrete values (quantization), objects have characteristics of both particles and waves (wave-particle duality), and there are limits to the precision with which quantities can be known (uncertainty principle). Naturally, scientists have tried to analyze consciousness with quantum mechanics.

On quantum consciousness, Penrose and Hameroff proposed a model called orchestrated objective reduction (Orch OR model). They suggested that quantum vibrational computations in microtubules were "orchestrated" (Orch) by synaptic inputs and memory stored in microtubules and terminated by Penrose's "objective reduction" (OR), hence "Orch OR." Penrose postulates that each separated quantum superposition has its own piece of space-time curvature, a blister in space-time. Penrose suggests that gravity exerts a force on these space-time blisters, which become unstable above the Planck scale of 10^{-35} m and collapse to just one of the possible states of the particle. The rough threshold for Orch OR is given by Penrose's indeterminacy principle.

Stapp envisages consciousness as exercising top-level control over neural excitation in the brain. Quantum brain events are suggested to occur at the whole brain level rather than the level of the synapses. Das proposes consciousness as the collapse of the interference pattern created by the Yukawa coupling of scalar fields of Nambu-Goldstone bosons and Dirac fields of electrons. The electrons come from the axons of neurons.

So far, it seems that no theory of the brain can explain why and how consciousness appears, if it assumes that consciousness is created by some neural entity that is totally different from behavior, structure, and function from our feelings. The only way out of this is to accept that consciousness must be a physical property. But physical property of what? The logical way is to analyze why consciousness is limited to the brain and which function and property of brain generates consciousness? Good age-old question!

From measurement of brain waves, we found that level of consciousness depends on the frequency of brain waves. When the brain is very active, consciousness is high and the brain wave frequency is in the gamma range (40–100 Hz). When the brain is active, consciousness is in beta range (12–40 Hz). In normal active stage, brain frequency is in the alpha range (8–12 Hz) and consciousness is medium. In the relaxed state, brain frequency is in the theta range (4–8 Hz) and consciousness is low. In the deeply relaxed state or dreamless sleep, the brain frequency is delta range (0.5–4 Hz) and consciousness is very low. Since consciousness is related to brain frequency, I have proposed that consciousness energy unit per brain wave frequency f for unit amplitude is $C.f$, where C is the consciousness constant *Conscire C*.

Total amplitude of a brain wave also is a factor of the square of the amplitude A of the wave, hence the energy of a brain wave at certain frequency $f = A^2.C.f$. However, when a brain wave is plotted by various techniques—for example, EEG, PET, fMRI, and so on—it is a combination of brain waves of various frequencies and amplitudes, depending on the mental state. The plotted brain waves could be analyzed with fast Fourier transform (FFT) or discrete Fourier transform (DFT), which will break down the complex waveform into discrete waves, each with its frequency and amplitude. However, the brain energy is very low; so far, it has not been possible to measure energy. As the technology keeps improving, it would be possible to measure the brain energy, hence find out the value of consciousness constant *Conscire C* from Eqn. 8.1 from the brain wave frequencies and amplitudes from DFT analysis of the brain waves.

APPENDIX 1

Zero-point energy

Zero-point energy is the energy of a system at absolute zero or the lowest quantized energy level of a quantum mechanical system. The origin of zero-point energy is the Heisenberg uncertainty principle, which reflects an intrinsic quantum fuzziness from the wave nature of the quantum fields. Planck proposed his second quantum theory, in which he introduced the zero-point energy. He found that the average energy ϵ of an oscillator is $\epsilon = h\nu/2 + h\nu/(e^{h\nu/kT} - 1)$ Eqn. (1).

h = Planck constant; ν = frequency; k = Boltzmann constant; T = absolute temperature. It can be seen from Eqn. (1) that at $T = 0$, there is still a residual energy $h\nu/2$, which is the zero-point energy.

Liquid helium-4 is a great example of zero-point energy. Under atmospheric pressure, even at near absolute zero, it does not freeze solid and is still a liquid. This is because its zero-point energy is large enough to keep it as a liquid. Casimir effect is another example of zero-point energy. The Casimir effect is a small attractive force that acts between two close parallel uncharged conducting plates. In quantum field theory, a vacuum is full of fluctuating electromagnetic waves in all possible wavelengths up to Planck frequencies. The cavity between the plates cannot sustain all frequencies of the electromagnetic field. Wavelengths comparable to the plate separation and longer are excluded from the regions between the plates. This leads to the situation that a zero-point radiation overpressure outside the plates acts to push the plates together.

Although there is no applied electromagnetic field, the two plates attract each other, the pressure being more and more as they move closer, just as two objects held together by a stretched spring moving closer as the energy stored in the spring decreases. S. Lamoreux at the University of Washington verified Casimir effect to within 5 percent in the size of a few microns. U. Mohideen at the University of California verified the Casimir effect even more precisely.

APPENDIX 2

Name	Symbol	Value
Speed of light	c	2.99792458×10^{8} m/s
Planck constant	h	$6.6260755 \times 10^{-34}$ J.s
Planck constant	h	$4.1356692 \times 10^{-15}$ eV.s
Gravitation constant	G	6.67259×10^{-11} $m^{3}.kg^{-1}.s^{-2}$
Boltzmann constant	k	1.380658×10^{-23} J/K
Boltzmann constant	k	8.617385×10^{-5} eV/K
Molar gas constant	R	8.314510 J/mol.K
Avogadro's number	N_A	6.0221×10^{23} mol^{-1}
Charge of electron	e	$1.60217733 \times 10^{-19}$ C
Permeability of vacuum	μ_0	$4\pi \times 10^{-7}$ N / A^2
Permittivity of vacuum	ε_0	$8.854187817 \times 10^{-12}$ F/m
Coulomb constant	$1/4\,\pi\,\varepsilon_0 = K$	8.987552×10^{9} $N.m^2/C^2$
Faraday constant	F	96485.309 C/mol
Mass of electron	m_e	$9.1093897 \times 10^{-31}$ kg
Mass of electron	m_e	0.51099906 MeV/c^2
Mass of proton	m_p	$1.6726231 \times 10^{-27}$ kg
Mass of proton	m_p	938.27231 MeV/c^2
Mass of neutron	m_n	$1.6749286 \times 10^{-27}$ kg
Mass of neutron	m_n	939.56563 MeV/c^2
Atomic mass unit	u	$1.6605402 \times 10^{-27}$ kg

Atomic mass unit	u	931.49432 MeV/c^2
Avogadro's number	N_A	6.0221367 x 10^{23} / mol
Stefan-Boltzmann constant	σ	5.67051 x 10^{-8} W/m^2. K^4
Rydberg constant	R_∞	10973731.534 m^{-1}

GLOSSARY

Abhidharma—Oldest Buddhist philosophy.

Abraham—First patriarch of Jewish people.

Agrarianism—A social philosophy that values rural society superior to urban society.

Altruism—Selfless concern for the well-being of others.

Alzheimer's disease—Alzheimer's is a type of dementia that causes problems with memory, thinking, and behavior.

Aristotle—Ancient Greek philosopher and student of Plato.

Artificial intelligence (AI)—An area of computer science that emphasizes the creation of intelligent machines that work and react like humans.

Atman—In Hindu philosophy, atman means inner self or soul. When a person achieves liberation or Moksha, atman returns to Brahman.

Axon—Long, slender projection of a neuron.

Baryonic—Concerning baryons consisting of protons and neutrons.

Bernard Baars—Neurobiologist at the Neurosciences Institute in California.

Bertrand Russell—British philosopher, logician, and mathematician.

Big bang theory—The universe was created from a point of infinite energy called singularity and has been growing since then.

Blackbody radiation—Blackbody is an object that absorbs all radiation falling on it. When a blackbody is at a uniform temperature, its emission has a characteristic frequency distribution that depends on the temperature. Its emission is called blackbody radiation.

Boson—Unlike electrons, bosons can group together. Photons making up light are bosons.

Brahman—In Hinduism, Brahman connotes the highest universal principle, the ultimate reality in the universe.

Brain stem—Area at the base of the brain that lies between the deep structures of the cerebral hemispheres and the cervical spinal cord.

Buddha—Gautama Buddha, or simply the Buddha, was an ascetic and sage on whose teachings Buddhism was founded.

Carl Jung—Swiss psychoanalyst and psychiatrist who founded analytical psychology.

Cerebellum—Sits below the cerebrum at the back of the skull.

Cerebral cortex—Thin layer of the brain that covers the outer portion of the cerebrum, often referred to as gray matter.

Cerebrum—Principal and most anterior part of the brain, located in the front area of the skull and consisting of two hemispheres, left and right, separated by a fissure.

Charles Babbage—English mathematician who originated the concept of a digital programmable computer.

Christof Koch—American neuroscientist best known for his work on the neural bases of consciousness.

Chronesthesia—Form of consciousness that allows people to think about something in subjective time and mentally travel in it.

Chit—Body and mind consciousness in Sanskrit.

Deepak Chopra—Deepak Chopra is an Indian American author, public speaker, and alternative medicine advocate.

Cognition—Mental action and process of acquiring knowledge or understanding through thought, experience, and senses.

Confucius—Ancient Chinese philosopher whose teachings founded Confucianism.

Cosmic microwave background (CMB)—Electromagnetic radiation left over from an early stage of the universe in big bang cosmology.

Cosmology—Science of origin and development of universe.

CT scan—Computerized tomography scan builds up a picture of the brain based on differential absorption of X-rays.

Dark energy—Unknown form of energy in the universe tending to accelerate the expansion of the universe.

Dark matter—Hypothetical type of matter distinct from baryonic matter (ordinary matter such as protons and neutrons) and neutrinos. Dark matter has never been directly observed; however, its existence would explain a number of otherwise puzzling astronomical observations.

David Bohm—American scientist who has been described as one of the most significant theoretical physicists of the twentieth century.

David Chalmers—Australian philosopher and cognitive scientist specializing in the areas of philosophy of mind. He proposes that consciousness is fundamental.

Decoupling—In cosmology, decoupling refers to a period in the development of the universe when the temperature dropped to three thousand kelvins and the photons were no longer able to disrupt the creation of the nucleus of an atom from proton and neutron.

Dendrites—Thin structures that arise from neurons.

Descartes—French philosopher often credited as the "father of modern philosophy."

DFT—Discrete Fourier transform allows Fourier transform to have a precisely defined range over which the transform will be calculated.

DICE model—American psychologist Daniel L. Schacter proposed his dissociable interactions and consciousness experience (DICE) model, which suggests that the primary role of consciousness is to mediate voluntary action under the control of an executive.

Dirac—Paul Dirac was an English theoretical physicist who made fundamental contributions to the early development of both quantum mechanics and quantum electrodynamics.

Doppler effect—Change in frequency or wavelength of a wave for an observer who is moving relative to the wave source.

Dualism—Believes that consciousness is separate from the body and has no relation to the brain.

EEG—Electroencephalography is the measurement of the electrical activity of the brain by recording from electrodes placed on the scalp.

Einstein—German-born theoretical physicist, considered one of the greatest physicists of all times.

Euclid—Greek Socratic philosopher who founded the Megarian school of philosophy. He was a pupil of Socrates.

Eugene Wigner—Hungarian-American theoretical physicist and mathematician.

Fourier transform—Breaks down time-based waveform into a series of sinusoidal waves, each with a unique magnitude, frequency, and phase.

fMRI—Functional magnetic resonance imaging, or functional MRI (fMRI), measures brain activity by detecting changes associated with blood flow. This technique relies on the fact that cerebral blood flow and neuronal activation in the brain are coupled.

fNIRS—Functional near-infrared spectroscopy is a noninvasive optical imaging technique using low levels of light to measure blood flow changes in the brain.

Francis Crick—British molecular biologist, biophysicist, and neuroscientist most noted for being a codiscoverer of the structure of the DNA molecule.

Friedrich Hans Beck—German physicist with research work on superconductivity, nuclear and elementary particle physics, relativistic quantum field theory, and theory of consciousness.

General theory of relativity—Einstein's general theory of relativity is one of the towering achievements of twentieth-century physics. It explains that the force of gravity arises from the curvature of space-time by mass, such as sun, Earth, and so forth.

Giulio Tononi—Neuroscientist and psychiatrist who is a distinguished chair in consciousness science at the University of Wisconsin.

Global workspace theory—A simple cognitive architecture proposed by Bernard Baars to account qualitatively for a large set of matched pairs of conscious and unconscious processes.

Goldstone—British theoretical physicist famous for the discovery of the Nambu–Goldstone boson.

Gordon Moore—Cofounder of Intel Corporation and the author of Moore's law.

Hameroff—Stuart Hameroff is an anesthesiologist known for his studies of consciousness.

Heisenberg—Werner Heisenberg, a German physicist, famous for his uncertainty principle in quantum mechanics.

Henry Stapp—American mathematical physicist known for his work in quantum mechanics and consciousness.

Higher-order monitoring theory—A mental state M of a subject S is conscious only when it has another mental state M*. It is the coming together of M and M* in the right way that yield consciousness.

Higher-order thought theory—A mental state is consciousness when its subject has a suitable higher-order thought about it.

Hippocrates—Greek physician considered one of the most influential figures in the history of medicines.

Homer—Greek author of two epic poems, *Iliad* and *Odyssey*.

Hubble—Edwin Hubble was an American astronomer. He played a crucial role in establishing the fields of astronomy and observational cosmology and is regarded as one of the most important astronomers of all time.

Hubble constant—Unit of measurement used to describe the expansion of the universe.

Immanuel Kant—German philosopher who is a central figure in modern philosophy.

Integrated information theory (IIT)—According to IIT, a system's consciousness is determined by its causal properties and is therefore an intrinsic, fundamental property of any physical system.

Isaac Newton—English mathematician, astronomer, and physicist who is widely recognized as one of the most influential scientists of all time and a key figure in the scientific revolution.

Jainism—An ancient religion originating in the Indian subcontinent, with a history of over three thousand years.

Jesus Christ—A Jewish preacher and religious leader. He is the central figure of Christianity. Christians believe him to be the Son of God.

John Eccles—Australian neurophysiologist and philosopher.

John Locke—English philosopher and physician commonly known as the "father of Liberalism."

Karl Pribram—Eminent American brain scientist, psychologist, and philosopher.

Legalism—A philosophical belief in ancient China that human beings are more inclined to do wrong than right because they are motivated entirely by self-interest.

Leibniz—German polymath and philosopher who occupies a prominent place in the history of mathematics and philosophy.

Leo Tolstoy—Russian writer regarded as one of the greatest author of all times.

Lepton—Electron with negative electric charge and neutrino with no electric charge.

Louis de Broglie—French physicist who postulated the wave nature of electrons and suggested that all matter has wave properties.

Max Planck—German physicist who originated quantum theory.

Maya—Illusion in Sanskrit.

McGinn's theory—Claims that an unknowable property P, a property of consciousness, somehow is responsible for the mind-body connection.

MEG—Magnetoencephalography is an imaging technique that identifies brain activity and measures magnetic fields produced in the brain.

Microtubules—Filamentous intracellular structures that are responsible for various kinds of movements in all eukaryotic cells (cells that contain a membrane-bound nucleus).

Mohism—Ancient Chinese philosophy in the teachings of Mo Di. It later merged into Taoism.

Monism—Believes that consciousness is part of brain activity.

Moore's law—Number of transistors in an integrated circuit doubles every two years.

Moses—A prophet in Judaism who led the Jews out of slavery in Egypt and led them to the Holy Land (the land of Israel).

MRI—Magnetic resonance imaging is a medical imaging technique to form pictures of the anatomy and physiological processes of body. MRI uses a powerful magnetic field, radio frequency pulses, and a computer to produce detailed pictures.

Muhammad—Last messenger of Allah (God) in Islam religion.

Nambu—Japanese-born American Yochiro Nambu is one of the most influential theoretical physicists of the twentieth century.

Nambu Goldstone bosons—Bosons that appear necessarily in models exhibiting spontaneous breakdown of continuous symmetries.

Neumann—John von Neumann was a Hungarian-American mathematician, physicist, and computer scientist.

Neuron—Nerve cell in brain.

Nick Herbert—An American physicist and author.

Niels Bohr—Danish physicist, regarded as one of the foremost physicists of the twentieth century, applied the quantum concept, which restricts the energy of a system to certain discrete values to solve the problem of atomic and molecular structure.

Objective time—Time measured by a clock and can be verified.

Occam's razor—A line of reasoning that says the simplest answer is often correct.

Orch OR—Orchestrated objective reduction. According to the general theory of relativity, mass is equivalent to space-time curvature. Objective reductions ("OR") are events which reconfigure the fine scale of space-time geometry.

Pali—Language of the scriptures of Buddhism.

Panpsychism—Consciousness, mind, and soul are universal and primordial features of all things.

Pauli—Austrian-born physicist Wolfgang Pauli discovered the Pauli exclusion principle, which states that in an atom, no two electrons can occupy the same quantum state simultaneously.

PANIC theory by Tye—Conscious experience should be poised, abstract, nonconceptual, and have intentional content.

Penrose—Roger Penrose is an English mathematical physicist, mathematician, and philosopher of science.

PET scan—Positron emission tomography scan uses radioactive substance called tracer to show how the brain and its tissues are working.

Phi phenomenon—Optical illusion of perceiving a series of still images, when viewed in rapid succession, as continuous motion.

Physical constant—Physical quantity that is universal in nature and constant in value.

Plato—Ancient Greek philosopher and student of Socrates.

Planck's constant—Planck solved the blackbody radiation problem by introducing Planck's constant, which states that the energy E of a light particle photon is given by $E = h\nu$, where h = Planck's constant and ν = frequency.

Prajna—In Buddhism, it is the wisdom that gives insight in the true nature of reality.

Prakriti—Nature or prime natural energy.

Purusha—The cosmic man or self, consciousness and universal principle.

Pythagoras—Ancient Greek philosopher and mathematician.

Quantum brain dynamics (QBD)—In neuroscience, QBD is a hypothesis to explain the function of the brain within the framework of quantum field theory.

Quantum computer—Computation system that makes direct use of quantum-mechanical phenomena, such as superposition and entanglement.

Quantum consciousness—The idea that consciousness requires quantum processes as opposed to the view of mainstream neurobiology, in which the function of the brain is wholly classical and quantum processes play no computational role.

Quantum field theory (QFT)—A field theory that incorporates quantum mechanics and the principles of the theory of relativity.

Quantum mechanics—Branch of physics that gives the fundamental theory of nature at the smallest scales of energy levels of atoms and subatomic particles.

Quantum number—Four quantum numbers are used to describe completely the movement and trajectories of each electron within an atom. Each electron in an atom has a unique set of quantum numbers; according to the Pauli exclusion principle, no two electrons can share the same combination of four quantum numbers.

Radhakrishnan—Indian philosopher, statesman, and second president of India.

Richard Feynman—American theoretical physicist known for his work in the path integral formulation of quantum mechanics and the theory of quantum electrodynamics.

Rousseau—French philosopher, writer, and composer.

Russellian monism—Conscious states are the intrinsic nature of brain states.

Samadhi—Meditative absorption or trance attained by the practice of meditation.

Schizophrenia—A mental disorder characterized by abnormal social behavior and failure to understand what is real. Common symptoms include false beliefs, unclear or confused thinking, hearing voices that others do not hear, reduced social engagement and emotional expression, and a lack of motivation.

Schrödinger—Erwin Schrödinger, Austrian theoretical physicist, contributed to the wave theory of matter and to other fundamentals of quantum mechanics.

Sigmund Freud—Austrian neurologist and founder of psychoanalysis.

Sikhism—Indian religion that follows the writings and teachings of ten Sikh Gurus, including the tenth Guru Gobind Singh.

Socrates—Ancient Greek philosopher credited as one of the founders of Western philosophy.

Special theory of relativity—Einstein proposed in this theory that nothing can exceed the speed of light. As the speed of an object approaches the speed of light, (1) time dilates (i.e., the clock slows down relative to a stationary one), (2) length contracts, and (3) mass increases significantly. This led to his famous equation, E = mc2, where E = energy, m = mass, and c = velocity of light.

Steady state theory—The universe has average properties that are constant in space and time so that new matter must be continuously and spontaneously created to maintain average densities as the universe expands. The universe has no beginning and no end.

Stephen Hawking—English theoretical physicist, cosmologist, and author.

Subjective time—Time based on personal view.

Swami Vivekananda—India Hindu monk who was a key figure in the introduction of Indian philosophies of Vedanta and Yoga to the Western world.

Synapse—The junction across which a nerve impulse passes from an axon terminal of one neuron to another neuron.

Taoism—Chinese religious and philosophical tradition.

Thalamus—Large dual-lobed mass of gray matter buried under the cerebral cortex. It is involved in sensory perception.

Thomas Aquinas—Italian Catholic priest, philosopher, and theologian.

Tubulin—Intracellular cylindrical filamentous structure and major building block of microtubules.

Turing machine—Abstract mathematical model of computation invented by English mathematician Alan Turing.

Umezawa—Hiroomi Umezawa was a physicist known for his fundamental contributions to quantum field theory and for his work on quantum phenomena in relation to the mind.

Upanishads—The Upanishads are a collection of texts of religious and philosophical nature, written in India probably between 800 BCE and 500 BCE, marking a transition from Vedic ritualism to new ideas and institutions. Although there are two hundred Upanishads, only fourteen are considered most important.

Vedanta—Literally means "end of the Vedas," reflecting ideas that emerged from the philosophies contained in the Upanishads.

Vedas—Collection of hymns and other religious texts composed in India between about 1500 BCE and 1000 BCE. The language of the Vedas is Sanskrit. There are four Vedas: the Rigveda, the Samaveda, the Yajurveda, and the Atharvaveda.

Vijnana—Our relative knowledge in Buddhism in which subject and object are distinguishable.

Voltaire—French writer, historian, and philosopher.

Yukawa—Hideki Yukawa was a Japanese theoretical physicist.

Zero-point energy—Energy of a system at absolute zero temperature.

REFERENCES

1. Gavin Flood, *An Introduction to Hinduism* (Cambridge University Press, 1996).
2. Swami Bhaskarananda, *The Essentials of Hinduism* (Seattle, 1994).
3. Paul Deussen and A. S. Geden, *The Philosophy of the Upanishads* (Cosimo Classics, 2010), 86.
4. *The Connected Discourses of the Buddha: A Translation of the Samyutta Nikaya* (Part 4 is "The Book of the Six Sense Bases [Salayatanavagga]") (Boston: Wisdom Publications).
5. "Modern Physics and Hindu Philosophy," http://www.grahamhancock.com/vasavadak2/.
6. Ronnie Littlejohn, *Confucianism: An Introduction* (New York: I.B. Tauris, 2010).
7. Xinzhong Yao, *An Introduction to Confucianism* (New York: Cambridge University Press, 2000).
8. http://www.en.wikipedia.org/wiki/Taoism.
9. Hugh H. Benson, ed., *Essays on the Philosophy of Socrates* (New York: Oxford University Press, 1992).
10. http://www.philosophybasics.com/philosophers_pythagoras.html.
11. Richard Kraut, ed., *The Cambridge Companion to Plato* (Cambridge: Cambridge University Press, 1992).
12. J. L. Ackrill, *Aristotle the Philosopher* (Oxford and New York: Oxford University Press, 1981).
13. Georges Dicker, *Descartes: An Analytical and Historical Introduction* (Oxford University Press, 1993).

14. Aidan Nichols, *Discovering Aquinas: An Introduction to His Life, Work, and Influence* (Grand Rapids, Michigan: Wm. B. Eerdmans Publishing Company, 2003).
15. V. S. De Laszlo, ed., *The Basic Writings of C. G. Jung* (The Modern Library, 1959).
16. http://www.journalpsyche.org/understanding-the-human-mind/.
17. P. A. Schilpp, ed., *The Philosophy of Bertrand Russell* (Evanston and Chicago: Northwestern University, 1944).
18. https://www.plato.stanford.edu/entries/voltaire/.
19. A. Rupert Hall, *Isaac Newton: Adventurer in Thought* (Oxford: Blackwell Publishers, 1992).
20. http://www.philosophybasics.com/philosophers_rousseau.html.
21. Sarvepalli Radhakrishnan and Charles Moore, eds., *A Source Book in Indian Philosophy* (Princeton: Princeton University Press, 1989).
22. P. Anstey, *John Locke and Natural Philosophy* (Oxford: Oxford University Press, 2011).
23. https://www.plato.stanford.edu/entries/kant-mind/.
24. Rita Carter, *The Human Brain.*
25. http://www.stephen-knapp.com/consciousness_the_symptom_of_the_soul.html.
26. https://www.theatlantic.com/magazine/archive/2015/04/the-science-of-near-death- experiences/386231/.
27. Swami Akhilananda, *Hindu Psychology: Its Meaning for the West* (London: Routledge & Kegan Paul Ltd, 1960), 26.
28. Ken Wilber, *The Spectrum of Consciousness.*
29. http://www.plato.stanford.edu/entries/consciousness/.
30. Robin Collins, "Modern Physics and the Energy Conservation Objection to Mind-Body Dualism," *The American Philosophical Quarterly* 45, no. 1 (January 2008).
31. Philip David Zelazo, Morris Moscovitch, and Evan Thompson, *The Cambridge Handbook of Consciousness* (Cambridge University Press).
32. "Altera's 30 billion transistor FPGA." Gazettabyte. June 28, 2015. Retrieved June 1, 2016.

33. "Quantum Computing," Stanford Encyclopedia of Philosophy, February 26, 2007, http://www.plato.stanford.edu/entries/qt-quantcomp.
34. http://www.spectrum.ieee.org/biomedical/imaging/can-machines-be-conscious.
35. http://www.scholarpedia.org/article/Integrated_information_theory.
36. http://ojs.concordia.ca/index.php/gnosis/article/download/143/107.
37. Colin McGinn, *The Problem of Consciousness* (Oxford: Blackwell Publishers, 1991).
38. http://www.ted.com/talks/david_chalmers_how_do_you_explain_consciousness/.
39. http:// www.plato.stanford.edu/entries/consciousness-higher/.
40. Bernard Baars, *A Cognitive Theory of Consciousness* (Cambridge University Press, 1998).
41. Philip Johnson-Laird, *Mental Models* (Lawrence Erlbaum Associates).
42. http://www.pdfs.semanticscholar.org/ae71/2a96ecdd13fe0e91e3cf23851cc4bf4eb613.pdf.
43. John Searle, *Mind, Language, and Society* (London: Weidenfield & Nicolson, 1999).
44. S. Dehaene and L. Naccache, *Cognition* 79, no. 1–2 (2001):1–37.
45. Geraint Rees, Gabriel Kreiman, and Christof Koch, "Neural Correlates of Consciousness in Humans," *Nature Reviews Neuroscience* (April 2002).
46. Tapan Das, *Why Astrology Is Science: Five Good Reasons* (Bloomington, Indiana: iUniverse).
47. John Von Neumann, *Mathematical Foundations of Quantum Mechanics* (Princeton University Press).
48. S. Hameroff and R. Penrose, *Toward a Science of Consciousness* (Cambridge: MIT Press, 1996), 47.
49. S. Hameroff and R. Penrose, "Conscious Events as Orchestrated Space-Time Selections," *NeuroQuantology* (2003).

50. S. Hameroff and R. Penrose, "Consciousness in the Universe: A Review of the 'Orch OR' Theory," *Physics of Life Reviews* (2014).
51. S. Rasmussen et al., "Computational Connectionism within Neurons: A Model of Cytoskeletal Automata Subserving Neural Networks," *Physica D42* (1990).
52. http://www.elsevier.com/about/press-releases/research-and-journals/discovery-of-quantum-vibrations-in-microtubules-inside-brain-neurons-corroborates-controversial-20-year-old-theory-of-consciousness.
53. http://www.scienceandnonduality.com/david-bohm -implicate-order-and- holomovement/.
54. J. Marciak-Kozlowska and M. Kozlowski, "On the Brain and Cosmic Background Photons," *NeuroQuantology* 11, no. 2 (June 2013):223–226.
55. A. Zee, *Quantum Field Theory in a Nutshell* (Princeton University Press, 2010).
56. C. I. J. M. Stuart, Y. Takahashi, and H. Umezawa, "On the Stability and Nonlocal Properties of Memory, *J. Theor. Biol.*, no. 71 (1978):605–618.
57. C. I. J. M., Stuart, Y. Takahashi, and H. Umezawa. "Mixed-System Brain Dynamics: Neural Memory as a Macroscopic Orders State," *Found. Phys.*, no. 9 (1979):301–327.
58. E. Del Giudice et al., "Electromagnetic Field and Spontaneous Symmetry Breaking in Biological Matter," *Nuclear Physics.* B275 (1983):185–89.
59. Henry P. Strapp, *Mind, Matter, and Quantum Mechanics* (Springer, 2003).
60. Y. Akbar, S. N. Khotimah, and F. Haryanto, "Spectral and Brain Mapping Analysis of EEG Based on Pwelch in Schizophrenic Patients," *Journal of Physics: Conference Series, Volume 694, conference 1.*
61. F. Beck and J. C. Eccles, "Quantum Aspects of Brain Activity and the Role of Consciousness," *Proceedings of the National Academy of Sciences of the United States of America* 89, no. 23 (December 1, 1992).

62. Tapan Das, "Theory of Consciousness," *NeuroQuantology* 7, no. 2 (2009):336–337.
63. Tapan Das, "Origin and Storage of Consciousness," *NeuroQuantology* 13, no. 1 (2015):108–10.
64. S. Hameroff and R. Penrose, "Orchestrated Reduction of Quantum Coherence in Brain Microtubules: A Model for Consciousness," *Journal for Consciousness Studies* (1996).
65. Tapan Das, "Origin of Consciousness and Zero-Point Field," *NeuroQuantology* 14, no. 1 (March 2016).
66. Thomas S. Kuhn, *Black-Body Theory and the Quantum Discontinuity, 1894–1912* (University of Chicago Press).
67. P. Baksa, https://www.huffingtonpost.com/peter-baksa/zero-point-field_b_913831.html
68. Daniel Bor, *The Ravenous Brain* (Basic Books).
69. Zero-point energy, http://www.calphysics.org/zpe.html.
70. S. K. Lamoreaux, "Demonstration of the Casimir Force in the 0.6 to 6 μm Range," Physical Review Letters (1997), Volume 78, Number 1.
71. U. Mohideen and Roy Anushree, "Precision Measurement of the Casimir Force from 0.1 to 0.9 μm," Physical Review Letters (1998), 81, 4549.
72. F. Crick and C. Koch, "Toward a Neurobiological Theory of Consciousness," Seminars in Neuroscience (1990):263–75.
73. G. Edelman and G. Tononi, *Reentry and the Dynamic Core: Neural Correlates of Conscious Experience* (Cambridge: Bradford Book, MIT Press 2000).
74. H. Flohr, "An Information Processing Theory of Anesthesia," *Neuropsychologia* 33 (1995):1169–80.
75. A. Damasio, *The Feeling of What Happens: Body and Emotion in the Making of Consciousness* (New York: Harcourt Brace, 1999).
76. Borjigin et al., "Surge of Neurophysiological Coherence and Connectivity in the Dying Brain," Proceedings of the National Academy of Sciences (2013):110(35):14432–14437.
77. http://www.hyperphysics.phy-astr.gsu.edu/hbase/Tables/funcon.html.

78. https://www.dataq.com/data-acquisition/general-education-tutorials/fft-fast-fourier- transform-waveform-analysis.html.
79. https//www.engineering.mit.edu/engage/ask-an-engineer/can-brain-waves-interfere-with-waves.
80. Tapan Das, "Origin of Singularity in Big Bang Theory from Zero-Point Energy," *Can. J. Phys.* 00: 1-3 (0000) dx.doi.org/10-1139/cjp-2017-0015.
81. http://www.chopra.com/articles/what-is-cosmic-onsciousness #sm.00018ep7zo7 meeeugh2cx267twx3.
82. David J. Chalmers, "Panpsychism and Panprotopsychism," in Alter and Nagasawa (2015):246–276.
83. http://www.plato.stanford.edu/entries/panpsychism/.
84. Bertrand Russell, *The Analysis of Mind* (London: George Allen and Unwin., 1921).
85. Bertrand Russell, *The Analysis of Matter* (London: George Allen and Unwin., 1927).
86. Paul Davies, *About Time*, (New York: Simon & Schuster, 1996).
87. http://www.quantamagazine.org/new-clues-to-how -the-brain-maps-time-20160126/.
88. http://www.people.vcu.edu/~dfranks/subjectiveand objectivetime.
89. http://www.drjoedispenza.com/files/understanding-brainwaves_white_paper.pdf.
90. Frank Amthor, *Neuroscience* (Hoboken, New Jersey: John Wiley & Sons, Inc.).
91. Tapan Das, "Consciousness as a Function of Brain Waves and Physical Constant Conscire," *NeuroQuantology* 15, no. 3 (September 2017).
92. Stanislas Dehaene, *Consciousness and the Brain* (New York: Penguin Group).

ABOUT THE AUTHOR

Tapan Das has an MSc in electronics from the University of London and a PhD in microwave electronics from the University of Bradford, England. He has thirty-eight years of experience in Plessey in UK, Nortel in Canada and US, Lucent Technologies in US, and SGNTT in Canada, starting from an engineer and leading to a vice president position working in the field of microchip design, television, telecommunications, wireless, and internet. The cofounder of Dastalk Telecom, he designed a call blocker to block telemarketers and unwanted calls. He has ten patents and has published many technical papers in international journals and conferences. He has written a book titled *Why Astrology Is Science: Five Good Reasons*. He is currently a director of the Ottawa Professional Engineers Ontario Chapter (PEO).

Tapan was awarded the Order of Honour from PEO and Fellowship of Engineers Canada for his outstanding contribution to the engineering profession. He has been researching consciousness and cosmology for the last ten years and has several published articles in international journals.

Printed in the United States
By Bookmasters